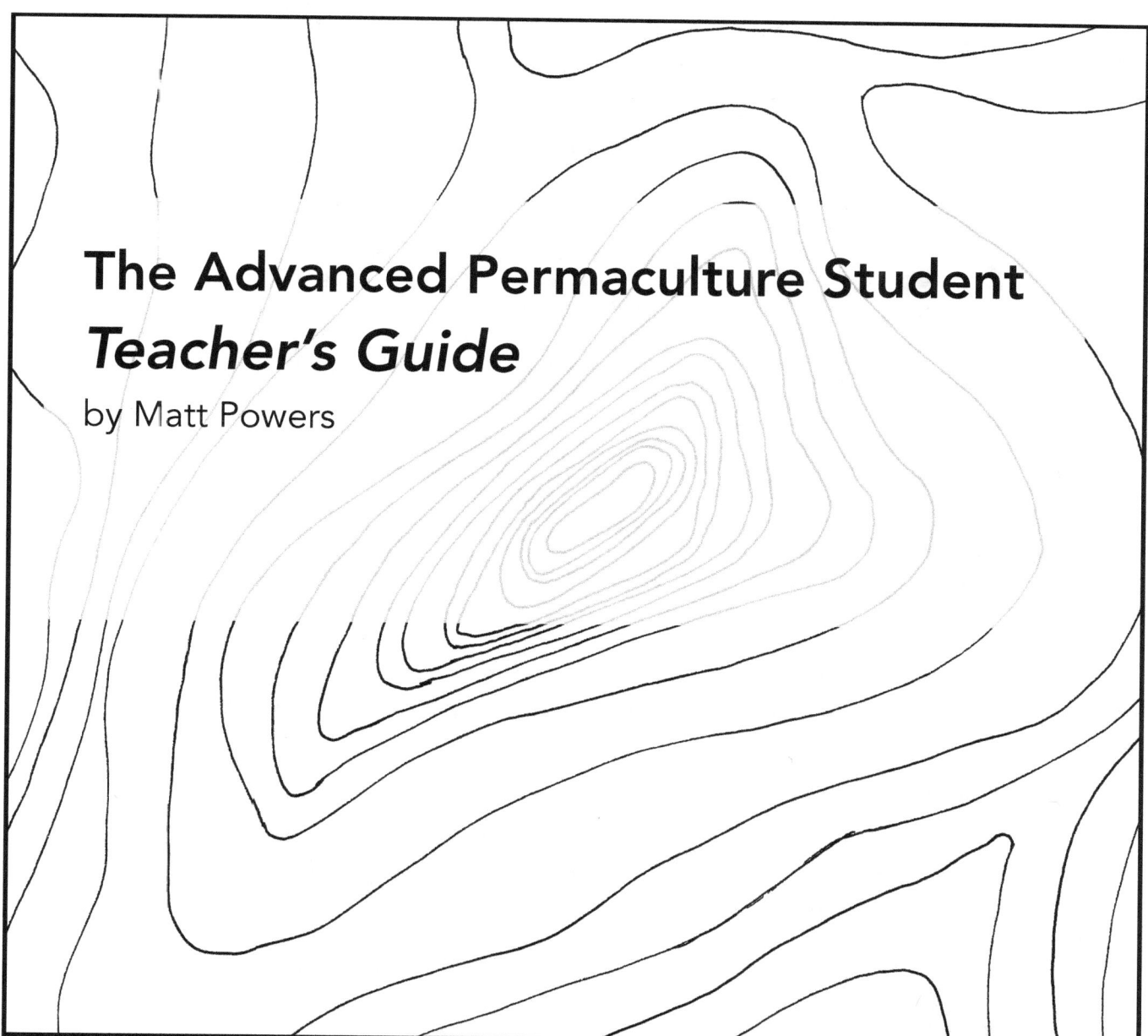

The Advanced Permaculture Student
Teacher's Guide
by Matt Powers

This Book is Dedicated to the 7th Generation

ISBN 978-1-7321878-5-6
Copyright 2018 All Rights Reserved by Matt Powers
Published by PowersPermaculture123/Matt Powers
Printed via IngramSpark.

Disclaimer:
Please Note: Just as food that nourishes one person causes an allergic reaction in another, the same concept of situational complexity applies to soils, dams, medicine, mushrooms, and more. In permaculture, complexity is embraced with the understanding that every situation and biome is unique. The information in this book represents research from sources listed—it is an educational and informational resource and does not represent any agreement, guarantee, or promise by any party associated with the creation or editing of this book. The publisher, editors, and author are not responsible for any negative or unintended consequences from applying or misapplying any of the information in this book.

Table of Contents

Introduction (1)
I. The Foundation for All Education (3)
II. Ethics, Principles, & Best Practices (6)
III. The Permaculture Education Standards (19)
IV. Lesson Plans & Project Proposal Samples (32)
V. The Future (82)
About the Author (83)

Introduction

Before I started on the collection of textbooks, workbooks, courses, and teachers guides that define my work in permaculture education, there was nothing available to teach children permaculture. There were some school gardening books for adults and some stories that could have been called permacultural but none that taught it directly. Neal Spackman of Sustainable Design Masterclass & the Al Baydha Project called it a "gaping niche", and I've been hard at work trying to fill it with everything I can to help showcase how large it is. So far, there has been an avalanche of attention in the elementary school space following my call and mission to create a K-12 curricular arc, but there have been no other attempts in the high school or middle school space – only individual programs at isolated schools. I'm hoping with the completion of the teacher's guides and the widespread adoption of permaculture standards as the core and foundation of all education K-12, we can see more textbooks, workbooks, and stories emerge to fill this critical gap in our culture.

This book is the guiding capstone on The Permaculture Student 2 book set, and if you want to teach an advanced permaculture design course, you'll need this book.

With this book, you can teach advanced permaculture – it includes lesson plans that pair with The Permaculture Student 2 and The Advanced Permaculture Student Online, the advanced permaculture certification program, but even more so, you can teach at an advanced level. This book covers best practices in education, advanced permaculture, social permaculture, and more – it also includes for the first time ever: Permaculture Educational Standards. It also has the current adopted standards, from the Next Generation Science Standards, that permaculture touches or expands upon.

This book is a keystone to the bridge between our education system and the regenerative economy. It's time to guarantee a brighter future for all by focusing on the foundation of our world's abundance: natural capital. If we can restore our vibrant ecosystems worldwide, we will regenerate the abundance that all human culture has always been based on. The only way to do that is by educating the next generation and thus guiding our entire culture towards that regenerative hope that we see glimmering through the permacultural lens.

To a noble, ethical, abundant, joyful, and thriving rewilded future!!

-Matt Powers

I. The Foundation for All Education

What Is Permaculture?

Permaculture is an ethical lens that uses nature's patterns and cycles to care for people, the Earth, and the future. It is finding balance and harmony with the natural world: a simple and ancient concept that is missing from modern daily life, and includes economic behavior, education, as well as entertainment. The abundance of the natural world is the foundation for all culture, economies, and life on Earth, so expanding and regenerating that foundation is the highest priority, the noblest calling, and the basis of all positive and productive interaction and education. Since education is how we shape our workforce, it is how we shape our economy. If we want a new economy, we have to embrace a new education system. If we want a regenerative economy, we must create a regenerative education system that guides a new generation to a new economy. Permaculture is the lens through which we can see that regenerative future.

There are many different words and names currently being used to describe partnering with nature. Why not use biomimicry? Why not agroecology? Why not agroforestry? While each of these concepts is powerful, vital, and compelling, they are part of a whole: Permaculture. The principles and ethics in permaculture make it unique – it helps guide interaction from micro to macro, personal to interpersonal, praise to discipline, privilege to responsibility, and further. Permaculture is a holistic concept that covers full spectrum of regenerative concepts, systems, and solutions. The organization, guiding ethics and principles, and holistic nature are what makes permaculture unique and the lynchpin in the regenerative movement and all future education.

Why Permaculture?

Permaculture is the core and foundation for all ethical education; it is interacting with other people, all forms of life, and the Earth with respect and care for the future. All variation is circumscribed by these boundaries. It is the birthright of every human being to know how to live regeneratively and ethically on Earth, so it should be the frame through which all education is viewed. To neglect or deny anyone the knowledge of how to support themselves without depleting or destroying their natural resources and to live ethically with other people and the living world is to deny them the most fundamental right of a living being.

Pattern literacy, a fundamental permaculture concept, is our most fundamental learning schema. The way we develop our understanding one subsection at a time to form a whole schema IS pattern literacy. We learn through patterns and then form them into holons (whole systems) or schema. The more developed our literacy, the more developed and detailed our schema: this is exactly how Olympic figure skaters coordinate innumerable body movements into a graceful and complete maneuver while skating on ice and leaping through the air. For children and young adults, being able to see the patterns within all systems, micro to macro, from the bioregional to the global whole is critical to giving students the deep analytical and observational skills needed to restore and regeneratively manage our world's fragile network of stressed ecosystems.

Permaculture involves and connects all disciplines: science, history, math, reading, writing, listening, speaking, design, engineering, physics, geography, geology, ecology, environmental science, organic chemistry, soil science, biology, agriculture, horticulture, animal husbandry, cooking, seed saving, art, problem solving, critical thinking, team work, reciprocity, compassion, and much more – it is everything within and touched upon by the 3 Ethics: Earth Care, People Care, & Future Care. It's everything we humans do that is positive and regenerative. It's how we grow and prepare our food – from seed to table back to soil again. It's the full cycle in every context. Every human being should know how to grow food, save seed, and prepare food as a basic building block for all of their education. In addition, every human being should have an education rooted in ethics that helps guide them on a path to a positive and abundant future for themselves and with everyone they interact with directly and indirectly. These ethics support and guide all the concepts that follow, so it's implicit in all aspects of life – even in discipline, governance, and currency.

Blending permaculture into lessons isn't hard, but it does require fluency in the topic you are teaching – you must know it well enough to explain it from different perspectives, so that it makes sense through multiple learning modalities and to multiple types of intelligences. It can be as simple as using the seeds the students just learned to save for a counting or basic arithmetic activity. It could be as advanced as designing and building working renewable energy, waste management, and water harvesting systems in high school and college programs. Math, science, engineering, electronics, physics, biology, microbiology, organic chemistry, and ecology can all be authentic, challenging, and rigorous; but when couched in engaging, student-centered, project-based, and hands-on learning experiences, the learning becomes supercharged and complexities and challenges are seen as fun and interesting. The permaculture standards, principles, and ethics will serve as the backbone and foundation for all education with all subjects and disciplines branching off of them. We can teach young children high level science, social skills, and regenerative concepts in a way that will reinforce their education and advance their progression, lengthening their strides of achievement and comprehension towards mastery in key life learning skills.

Can I Teach Permaculture? Do I Need A Certification?

Permaculture design certifications are increasingly common, easy to acquire, and are no indicator of level of skill or understanding. Many professional teachers who teach PDCs have never even acquired their own certification but instead were already doing regenerative work and their work alone made them an accepted expert. A PDC certification may be valuable for working with younger children in the K-5 space, but in middle school and high school that set of information should be seen as introductory only. The way permaculture certification was setup by Bill Mollison was to allow anyone to teach permaculture to anyone anywhere in the world but to require certification to be granted by those already certified through a 72-hour program with a design requirement on paper or digitally following the information set found in Permaculture: A Designer's Manual (1989). That information set is almost 30 years old now – much has changed in our understanding of our world in that timeframe. Learning from leading professional regenerative educators, sourcing peer-reviewed curricula, and earning your advanced permaculture certification are the best steps to take in your development as a permaculture educator and practitioner.

II. Ethics, Principles, Standards, & Best Practices

The Three Permaculture Ethics

Earth Care - to always protect, preserve, conserve, restore, and regenerate the Earth and all its inhabitants, elements, cycles, and systems in all ways

People Care - to always protect, preserve, shelter, nourish, heal, educate, inspire, and empower all people holistically

Future Care - to fully commit to consistent, long-term investments in longview planning for and long-term management and care of both natural and people systems

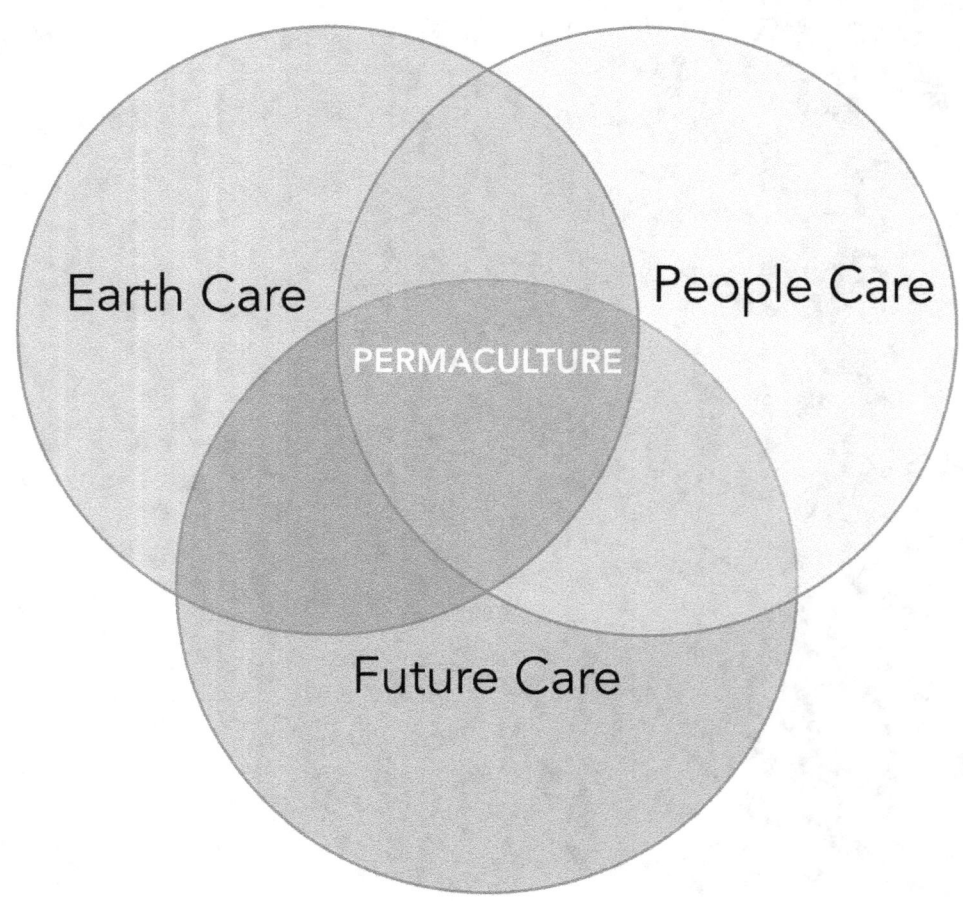

Permaculture Principles

- ***Principle of Observation*** - Observation and reflection are the greatest tools we have as humans. We can learn, adapt, and progress in our understanding limitlessly, but it can only be achieved through listening, observing, and pondering. We always run the risk of imposing solutions before we observe what the situation might actually be calling for. A lengthy observation is always better than a hasty action, especially when we are moving soil or building a home.
- ***Rule of Necessitous Use*** - We only use what we need, leaving nature alone whenever possible. We can already make sustainable living systems ethically on less land than we are currently utilizing.
- ***Principle of Cooperation*** - Cooperation is the name of the game in nature. It may be hard to see at times, but there is a constant exchange like a vibrant economy at work in our forests, fields, and soils. Everything works in cooperation: occupying niches, participating in cycles, balancing each other's populations, removing unwanted traits and trading inputs and outputs when seen from the perspective of energy, water, or the larger life cycles of that species or ecology.
- ***Life Intervention Principle*** - Chaos creates the greatest opportunity to implement creative design simply because as designers, we can create order, and not in some linear, straight-lined, or square fashion, but in nature's syntropic concept of order which usually looks wild and disordered to the eye. From the monoculture fields to the chaotic ecologies coping with climate change and invasive species, they all provide an opportunity for order through a biological intervention whether it be through a human, a beaver, or termites—life is the catalyst in creating order out of chaos.
- ***Principle of Return*** - As with the concept of whatever goes up must come down, this principle is about how every cycle and process in nature requires an input. Whatever we harvest, we must in turn sow. If we take, we must return.
- ***Syntropy Not Entropy*** - Entropy, the traditionally accepted idea that energy is always dissipating and that order is always headed towards disorder, is false in the sense that it overlooks how life systems work. Life is syntropic: it attempts to organize systems to trap and cycle energy in as many ways and as many times as possible while generating more and more life. All our systems and behaviors must be syntropic—creative designs that channel energy and generate more life.
- ***The Principle of Empowerment*** - The best natural systems are self-managed and need no human intervention. To relinquish power to the people, the land to the wildlife, to share

responsibility - it all empowers through decentralization and makes stronger, more resilient ecosystems, countries, companies, families, and neighborhoods.
- ***Principle of Purity and Preservation*** - No methods can be used that will degrade, taint, or completely consume any resource. We need to preserve our limited stores of freshwater and keep them pure. This same principle can be applied to our soil, air, bodies, cities, forests, and more.
- ***Principle of Choice*** - Dictating behavior or preventing natural behavior both prevent choice, and, in natural systems, prevent systems from thriving.
- ***Principle of Stability*** - Though more diversity is ideal, the mutualistic connections among elements in the ecosystem are the actual source of benefit. Everything interacts through the web of life and therefore holds intrinsic value, but certain associations are more beneficial than others, and we can select and combine them for greater stability.
- ***Principle of Self-Regulation*** - Self-regulating systems are more stable and long lasting than systems maintained by inputs and managed by people. This only happens when elements are aligned to work together.
- ***Definition of System Yield*** - The concept of yield is not limited to what we harvest from a single crop; it is the total energy produced, recycled, reused, converted, trapped, and conserved over a period of time—taking into account the energy consumed by the processes. Only when we see a holistic picture of energy-in and energy-out can we see how efficient and effective a system is ecologically.
- ***The Role of Life in Yields*** - All yields in nature come from biological processes. Without life, the sand, clay, and silt will never break down into soluble minerals and nutrients that plants and animals can absorb. The efficiency and diversity of the biology in a system determines how high yields will be.

Birch's Six Principles of Natural Systems

- ***Nothing in Nature Lasts Forever*** - Though some trees can grow for thousands of years, such as olive trees that were alive during Roman times, all life reaches its peak in growth, maintains for a time, and then eventually declines into decomposition.
- ***Natural Cycles Perpetuate All Life*** - In the web of life, all things interact directly and indirectly through localized and larger natural cycles, like the water cycle, the soil food web, and the seasons.
- ***Extinction Occurs With Very High or Very Low Populations*** - When vital resources become scarce in high population situations, extinctions can occur quickly, like when large schools of fish get caught in shallow, warm waters (that have low oxygen levels), and a

mass die-off occurs. There is also a minimum genetic diversity needed to prevent extinction. With corn, you typically need a minimum of 100-200 plants to maintain enough genetic diversity to maintain that line and prevent inbreeding depression which makes the corn ears and kernels shrink. Some corns, like Glass Gem and Painted Mountain corn, are genetically diverse enough that they show some resistance to inbreeding depression.

- **Every Species Has Key Elements That It Depends on to Survive** - When farmers that use **biocides** focus on killing a certain plant or bug, they tend to cause a chain reaction that threatens critical pollinators and other vital cycles. Unintended consequences come into play when we don't see the full extent of relationships in an ecosystem. That is why we work with nature, observe nature, and learn from nature, because those key elements that intertwine within the ecosystems we live in, we depend upon as well.
- **Our Ability to Change the Earth Always Precedes Our Ability to Foresee What the Consequences will be** - We make change faster than we can foresee results, so we must always observe, test, and plan carefully before we act. Always spend more time observing than in action.
- **All Life Has Intrinsic Worth** - Everything living has a function even if we cannot readily perceive what it is. Observation and respect is needed when encountering or interacting with all life. With perhaps the exception of some humans, all biodiversity is trying to participate in the cycles of life and to create more life.

Mollisonian Permaculture Principles

- **Work with Nature** - Instead of trying to control, pacify, "tame," or dominate nature, we need to work with nature and recognize the truth that we <u>are</u> nature, and our survival is dependent on our positive relationship with nature.
- **The Problem Is the Solution** - Perhaps the widest-spread permaculture principle, the concept that our problems are really indicators of solutions is a powerful one. It puts a positive spin on negative situations while focusing on the systems surrounding the "problem." For example, excess manure from farm animals can become a serious problem if allowed to accumulate and then leach into groundwater, but it can also be a high quality soil amendment if properly composted. The toxic situation could be growing us healthy food instead.
- **The Least Changes for the Longest Term Effects** - To maximize our effect and minimize our efforts, we need to seek the longest term and most ethical solution that requires the least amount of input energy and maintenance. When we route our washing machine's

graywater into our yard (while using safe, non-toxic detergent), we save an immense amount of water and irrigate our gardens routinely.
- **_Yields Are Limitless_** - Only our imagination limits our ability to find more yields from a finite system. It is often described using the example of how Native Americans traditionally used the entire deer, wasting nothing. This abundance and efficiency perspective leads to viewing the world differently. Yields are also limitless in the sense that through stacking functions, they can diversify nearly endlessly even in a finite system. Total yields must be accounted for not just individual yields.
- **_Everything Gardens_** - This can be easily observed in nature. The moles aerate the soils, redistribute seed, and fertilize it like tiny, blind subterranean farmers. Birds spread seeds to the edges of forests and expand the system, as do deer and many other animals. All these animals fertilize the soils around the plants they eat from. Even non-living elements, like the wind and water cycle, contribute to the spread of more life. Everything works together in a symphony of gardening and food forestry (even if it's not always food humans can eat).

Rules of Conservative Use

- **Reduce Waste and Pollution** - Everything is part of a cycle in process. All waste is a resource in the next step in a cycle or several cycles. It is our responsibility to cycle our waste safely and regeneratively; it helps to produce waste that can be cycled easily. Switching from plastic grocery bags to reusable, natural fiber bags is a simple, easy step that dramatically changes one's impact when the natural fiber bags break down in the compost heap in a few weeks—whereas the plastic bags can persist for decades.
- **_Restoration of Mineral and Nutrient Cycles_** - Over the past century of industrial farming, the perpetual tilling, expansion, and persistent use of salts to time-release fertilizers and biocides has left us with dirt: dead soils which are deficient in soluble nutrients and soil life. Through the return of biodiversity and the conservation of resources and ecologies, we can bring our soils back to life. We also need to turn around and mine the landfills we've created and begin to sort out all the resources we've mixed and to various degrees decomposed.
- **_Careful Energy Accounting_** - When someone buys solar panels, they often don't consider where the raw materials were harvested from, how they were harvested, how they were processed and transported, how those refined materials combined and were transported to you, and how much time and energy they will require to break down, including the panels themselves. Understanding the amount of energy it takes to complete these

processes and to cycle the waste shows us the true cost and nature of that product or process.
- **Identify and Prevent Potential Long-Term Negative Effects** - This may seem hard to do, but business plans in Japan traditionally looked 100 years into the future, so it's not unheard-of thinking. Generational thinking is needed and looking ahead for unforeseen problems is critical as we plan and implement our land restoration projects. We need to look at the 50- and 100-year storms and see if our dams will be able to handle that amount of water safely. If we don't look ahead, we will find ourselves unprepared.

Permaculture Ethics in Landscape and Society

- **Care for Natural Systems** - The remaining pristine wilderness must be protected to prevent further habitat destruction and to preserve the natural systems that provide our air, soil, and water.
- **Rehabilitate Degraded or Damaged Ecosystems** - Most of the agricultural land used in the past 10,000 years is now infertile. Most of the primeval forests that dominated the landscapes of the globe are gone. In some areas we will be able to support and restore the struggling ecosystems while in other areas we may be reintroducing vegetation.
- **Create Our Own Beneficial Living Systems** - Using permaculture, we can create our own complex environments to support ourselves. Every homestead can be a beneficial microclimate of human habitation. When we take care of ourselves, we remove the pressure we are putting on the planet elsewhere.

Social Permaculture Principles

- **Treat Others Better than They Expect to Be Treated** - If we want people to be willing to dig swales, grow their own food, etc., we have to make it a pleasant experience. If we treat others better than they expect, they will be more trusting and open.
- **Show Trust and Be Trustworthy** - The only way to gain trust is to show that we are worthy of it by demonstrating trust first and then being trustworthy.
- **Be Clear** - When we are clear, our intentions are understood, and others can meet expectations—which also builds trust.
- **Set Clear Boundaries** - Like all edges, boundaries are areas for productivity. When we set clear boundaries, it shapes expectations and guides behaviors towards mutually beneficial ends. It also protects us and allows for growth, reflection, privacy, individuality, autonomy, healthy relationships, and much more.

- **Educate by Example** - The only way to spread good examples of design or behavior is through living those behaviors or thriving inside those designs. People have to see it, touch it, taste it, experience it, and know the story of it to adopt a significant change in the way they live their lives.
- **Share as Much as You Can** - This may be difficult as we are starting out, but everything we design can turn quickly into abundance and enable us to return surplus and prepare for the future. It is also a main component of the Third Ethic.
- **Be Self-Reliant (Be Prepared)** - While seemingly in juxtaposition to the last principle, the first step to helping others is being able to help ourselves, especially in relation to unforeseen complications. Planning and preparation are vital components of self-reliance.
- **Be Patient** - A tree does not grow in a day nor a forest in a year. The best things take time and patience—especially with people.
- **Be Local** - When we focus our time, energy, spending, and production locally and regeneratively, we improve the local area and reap all the associated holistic benefits.
- **Be Open** - Trust in you or in your business can only happen if you are transparent and open about how you conduct business.
- **Be Timely** - Whether it's plant, animal, or people systems, timing is everything.
- **Solutions, Not Complaints** - Though critical thinking and pointing out places for improvement is vital for reflection, complaining and finding solutions are worlds apart. When we speak in the language of solutions, we are more likely to arrive at solutions.
- **Smile First** - Enthusiasm is the energy of will made manifest. We all want to be around someone who is enthusiastic and ready to work.
- **Family First** - Families are the most basic units of all communities, the foundation of all civilization, and the purpose for our lives.
- **Work First on What Matters Now** - Always ask yourself: What is important now? Always keep your priorities focused on what is needed now to accomplish your holistic goals.
- **Always Innovate and Adapt** - Nature is always innovating and adapting to the constantly changing environment. If we are to work with nature, we must change with it.
- **Don't Take Offense, Be Better** - Often it is the most difficult challenge, but, ultimately, the ability to consider criticism without taking offense brings the greatest set of socio-environmental rewards both individually and collectively.
- **Look to Elders** - Experience is priceless, and we all can learn from earlier generations' perspectives even if we choose to do something different. All perspectives play a part in holistic management.

- **Celebrate Common Interests** - When we meet someone who shares a passion we have, we cannot help but feel like we've gained a new friend and ally. A common bond can be found between almost any two people because we all have universal needs.
- **Listen to and Make Space for Children and Youth** - Children, youth, and young adults also tend to be marginalized, despite families having done so much in order to raise children; making room for the younger people to be heard and to sense themselves as autonomous beings is important.
- **Include Everyone When Possible** - Permanent cultures rely upon holistic representation and celebration of everyone, regardless of culture, gender, or beliefs. The cultural edges can bring a spectrum of experience and perspective that would be impossible otherwise. This might not always be possible — some issues might be personal, trivial, inappropriate, outside their concern, and a myriad of other possibilities.

Guiding Educational Principles & Concepts

Did you know there is a succession to thought? It takes more understanding and experience to create something than it does to recognize or define it. If you cannot whistle but are old enough to understand whistling and recognize it (identification), you know that it is blowing air through your lips in a very precise manner (comprehension). You struggle to apply it until you have a breakthrough (application) and you begin to analyze HOW you just did that and how it all works (analysis), where are the limitations of your skill, and then you try to synthesize your new skill with a novel setting: whistling a song you know (synthesis). As you listen to yourself perform a desired note or an entire song, the critical mind and faculty awaken and you ask: was that good? How did that sound? What can I do better or differently the next note or run-through (critical thinking)? Lastly students can chart their own course and create their own unique expression of that concept or using that skill (creation). Almost everywhere educators are trained, they learn about Bloom's Taxonomy of Cognition, so they can order their lessons to build in complexity in a logical succession of thought. It follows the way learners' thinking progresses.

Bloom's Taxonomy of Cognition

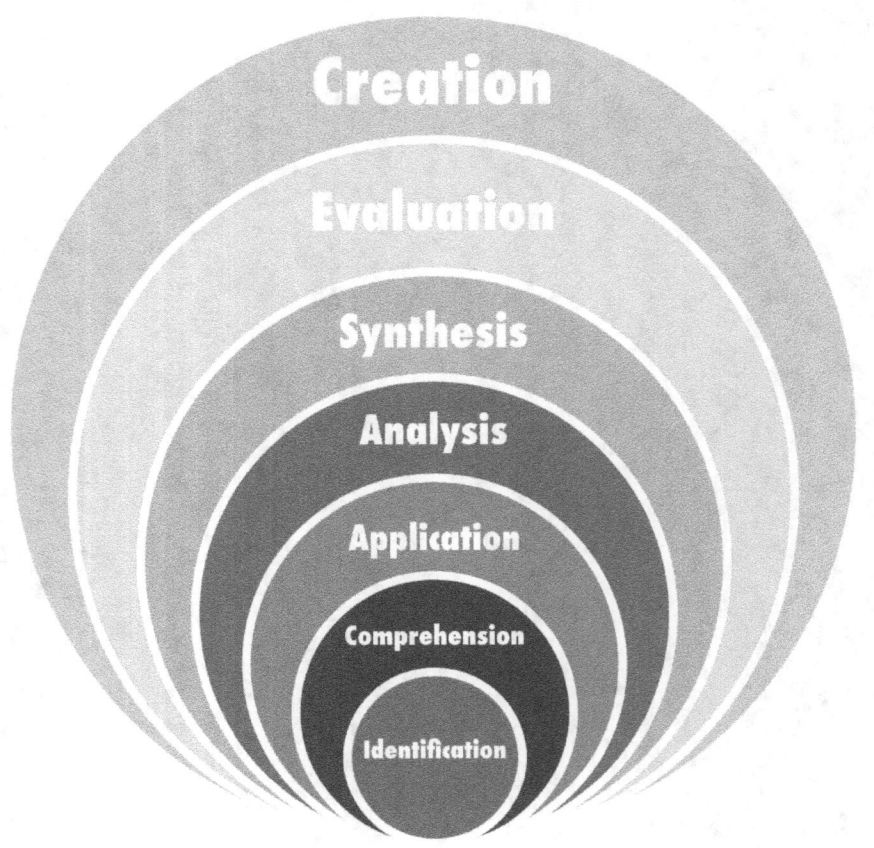

Did you know that everyone learns differently... in different contexts? You likely know folks that can read and retain information differently than when they are told verbally, and vice versa. Some students learn through movement, some through listening, some through seeing it done, and some from using their hands while they are watching you doing it – it all depends. If you look up learning modalities, you typically just find: auditory, kinesthetic, and visual, but there are more. People are more nuanced, and we all have varying degrees of ability within each of these modalities. In addition, we can grow and improve in each and in all as well. As educators, we can use these modalities to reach more students, to stretch them further, and to connect their prior learning schema to new learning and even to transfer skills – as is what happens when an avid audiobook listener who is also learning to read transfers their listening comprehension skills to reading comprehension skills which at the same time leads to rapid improvements in reading retention, reading speeds, spelling, grammar, and writing skills.

Learning Modalities

- Auditory Learners
- Kinesthetic & Tactile Learners
- Visual & Spatial Learners
- Analytical Learners
- Holistic Learners
- Social & Verbal Learners
- Game & Competition Learners

Not only do we not learn in the same way, but we also have different intelligences, and these, too, can be strengthened through practice and study, but many of these intelligences are deeply ingrained and part of who we are as a person, thinker, and learner. This also forms subtle biases in us as educators – we praise and reward those who do well in our classrooms without recognizing that most succeed at a high level because they are already strong in our own particular form of intelligence. While we want those students to reach as high as possible in their own intelligence and modality, the students who are not already primed in that form of intelligence may be working harder to do satisfactory work in the same context. Every learner has their own unique context, prior learning, and strengths in a combination of intelligences though usually we are strong in one or two areas. Traditional schooling has focused on math/logical and verbal/linguistic for standardized testing and grading while ignoring the many other types of ways we learn, think, and express ourselves.

9 Forms of Intelligence

- Nature
- Music
- Body/Kinesthetic
- Spatial/Visual
- Interpersonal (communication between people)
- Intrapersonal (metacognition, understanding thought)
- Math/Logical
- Verbal/Linguistic
- Spiritual/Philosophical/Existential

Permaculture Education Principles

- We can always learn something new from carefully observing natural systems
- We all can learn to be regenerative in all ways though it will require time, effort, and adaptation
- Learning cannot be forced, only shared and accepted. No one can force anyone to learn anything. We can only invite, present, and offer a pathway. All learning is acceptance of a truth and that is always a choice
- Everyone can learn, grow, and progress though it requires a growth mind-set to do so
- "Kids do well when they can" – Dr. Ross Greene. Students fail when they lack the skills, schema, or confidence (which comes from competence)
- Our problems can become our greatest solutions and can transform us if we have a growth mindset

- Teaching is the highest level of learning – have your students teach as much as possible
- Enthusiasm is the currency of a vibrant learning community, like sunlight to a forest – <u>Bring the Enthusiasm</u>
- Beyond Socratic is Heuristic – help students ask better questions by having them practice more often
- Reflection is key to deeper understanding, problem solving, and adaptation
- Authentic Permaculture Education leads to Regenerative Action & Empowerment
- Holistic empowerment is key to maintaining sustained growth
- Pairing regenerative practices with service to the community and those in need deeply encodes the learning and significance of that learning
- Self care is central to all care

Best Practices for Permaculture Educators

- Reach your students through their own modalities, prior knowledge, and experience
- Build trust in your classroom and community – it is the basis for the feeling of safety and comfort needed for deep learning
- Regularly connect new learning to permaculture principles and the 3 Ethics, especially when the focus becomes more technical, complex, or hands-on
- Use frameworks to organize thinking and designs, but be flexible and be ready to adapt, combine, and rethink
- Create lessons and projects with choices, like a menu, and present them as an invitation
- Create lessons and projects that rely upon, strengthen, and support all forms of intelligence not just math and verbal
- Allow students to express their learning in modalities they are comfortable in, but challenge your students to challenge themselves
- Use their holistic goals as the foundation for self-appointed challenges based in curiosity
- Everyone has a story – facilitate an environment and experience where all can feel they can share their own story
- Keep lessons, projects, and themes student-centered to maintain engagement and deep learning

- Use projects to bundle skills and learning into authentic learning experiences and performance assessments
- Invite students to take on the role of the teacher as a summative assessment of their learning
- Use groups to support peer-to-peer learning and sharing
- Model best practices for self-improvement and self-care: positive self-talk, stepping out problems with analysis and critical thinking, regular meditation, regular exercise, and healthy spiritual and social relationships. This does matter — when we present holistic health in an accessible, generalized manner, we help students find that balance as well even when it isn't modeled at home and even when it is nothing like what they believe or will do to maintain their holistic health
- Model and use nonviolent communication especially with students in conflict or crisis situations
- Use restorative justice practices like restorative circles to mediate conflict and to resolve conflicts
- Build a classroom community of growth mind-set life learners who see mistakes as part of the learning process and critical to growth
- Meet your students where they are at and at their level — this may mean at their eye level, sitting among the students instead of off to the side, or organizing the lesson around their interests and zone of proximal development
- Foster a community of lateral, cooperative, peer-to-peer learning and decision making by using questions and prompting students to take the lead asking questions and planning projects
- Build a community wherein social permaculture principles and nonviolent communication are practiced habit

III. The Permaculture Education Standards

Some standards could easily fall into two or more sections, but are not repeated in each section for the sake of brevity and clarity. Also these standards represent an end goal for each academic time period; therefore teachers are expected to preview concepts and skills before students are expected to understand, apply, analyze, or critique them. These standards are clear objectives but do not represent the breadth of information that can and should be covered in each academic time period, rather, they cover the student behaviors and experiences expected from the educational program.

Why include elementary & middle school standards? My experience as a teacher of hundreds of high school students is that there is a lack of cohesion and communication between grades, subjects, teachers, and administrations. Presenting the standards together as whole K-12 progression brings that coherence and iterative clarity. On top of that, most teachers will not have students that have had a K-8 permaculture-focused education, so the lower level standards and pathway are useful for scaffolding purposes. Lesson plans for these particular groups (Elementary & Middle School) are found each in their own teacher's guide where the standards are also presented as a K-12 full arc, so all teachers see the end goals and the outset as they teach their portion. Over time, these standards, too, will be adapted as natural pattern literacy becomes more common in homes and family life.

The Purpose of Permaculture Education Standards

Teachers need standards to guide and align their lessons and assessments — administrators need them as well to adequately support, critique, and guide teachers. Standards are the lynchpin and measuring rod for a given subject. Iterating a subject out over time allows it to be thoroughly taught, deeply understood, and, at the same time, communicated and taught within the zone of proximal development. For permaculture, it is no different: educators want clear standards and lesson plans aligned to cognitive and behavioral abilities and expectations that match the developmental stage of that age group.

Elementary
(Identification, Comprehension, & Basic Skill Acquisition)

1. Comprehension
1.1 Students are able to identify and define core concepts: the 3 Ethics and the definition of
permaculture
1.2 Students are able to identify and describe permaculture in context

2. Knowledge
2.1 Students are able to share multiple examples of permaculture in action
2.2 Students are able to identify and list multiple regenerative solutions for local energy, food, water, and waste management needs
2.3 Students are able to identify their bioregion and watershed
2.4 Students are able to identify 10-20+ local native plants, 5-10+ native pollinators, 5-10+ local fungi, and 10-20+ local animals as well as their roles and interactions in their bioregional ecosystem
2.5 Students are able to identify 20-40+ annual and perennial garden and food forest plants
2.6 Students are able to identify, create a representation of, and explain the water cycle, the
carbon cycle, the solar cycle, and the bioregional seasonal conditions

3. Design
3.1 Students are able to design a simple garden on paper
3.2 Students are able to design a simple home site on paper, including regenerative solutions for energy, food, water, waste, and shelter
3.3 Students are able to identify, create, and comprehend simple maps and designs

4. Regenerative Skills
4.1 Students are able to grow a diversity of plants from seed
4.2 Students are able to harvest and preserve seed for the next season
4.3 Students are able to harvest and prepare food they have grown or helped grow in a variety of ways, forming several complete meals from their garden or food forest
4.4 Students are able to assist in setting up and managing a garden
4.5 Students are able to preserve food in a variety of ways (canning, drying, pickling, etc.)
4.6 Students are able to make vermicompost (using worms) to process food waste

5. Social Skills
5.1 Students are able to identify, define, and share examples of compassion, empathy, and people care

5.2 Students have participated in service, connecting natural principles to social principles

5.3 Students are able to model nonviolent communication and restorative justice skills in everyday settings

6. Self Smart Skills

6.1 Students are able to write out their holistic goals

6.2 Students are able to meditate for 5-10 minutes at a time

6.3 Students are able to do rudimentary exercise and stretching, primarily through physical

   play, dance, and games as well as yoga, qigong, tai chi, and jogging

6.4 Students are able to develop healthy and balanced meal ideas based on bioregionally and homestead or school garden sourced foods

Middle School
(Identification, Comprehension, Application, Analysis, Synthesis, Intermediate Skill Acquisition, & Service)

1. Comprehension & Analysis
1.1 Students are able to identify, define, apply, and synthesize key concepts: the 3 Ethics, permaculture principles, social permaculture principles, and the definition of permaculture

1.2 Students are able to present and teach a variety of permaculture skills and concepts to others using multimedia, public speaking, social media, and other modalities

2. Knowledge
2.1 Students are able to identify and map their bioregion and watershed

2.2 Students are able to identify 20-50+ local native plants, 10-20+ native pollinators, 10-20+ local fungi, and 10-30+ local animals, their roles and interactions within the ecosystem as well as any related indigenous history and uses

2.3 Students are able to identify 50-100+ annual and perennial garden and food forest plants and their native origins and uses

2.4 Students are able to identify, grow, harvest, and cook edible and medicinal mushrooms
 in an outdoor setting (shiitake, oyster, reishi, etc.)

2.5 Students are able to identify, create a detailed representation of, and teach the water cycle, the mineral cycle, the carbon cycle, and the global annual seasonal cycle in relation to the sun

2.6 Students are able to identify and describe the different components of soil, the soil food
 web, and photosynthesis in relation to the soil food web

3. Design
3.1 Students are able to create a basic permaculture design for a home site

3.2 Students are able to design, set up, and manage a small garden or garden plot

4. Regenerative Skills
4.1 Students are able to problem solve using permaculture and explain why and how

4.2 Students are able to grow a broad diversity of plants from seed, cuttings, and tubers

4.3 Students are able to harvest and preserve a broad diversity of seed for long-term storage, forming their own seed bank

4.4 Students are able to prepare and preserve food they have grown or helped grow in a variety of ways, forming a diversity of meals and preserves that are shared, consumed with the class, taken home, or sold

4.5 Students are able to make both thermophilic (hot) compost and vermicompost (using worms) to process food waste

4.6 Students are able to cultivate aquatic plants and animals in a small controlled environment and observe and assist in cultivating aquatic plants and animals in a larger context

5. Social Skills

5.1 Students have participated in several permaculture service projects each year, pairing the social principles of permaculture with natural principles – providing a minimum of 15 hours of community service

5.2 Students are able to identify, define, share, plan, promote, and participate in actions of compassion, empathy, and people care

5.3 Students are able to model nonviolent communication and restorative justice skills in everyday and conflict mediation settings

6. Self Smart Skills

6.1 Students are able to write out their holistic goals

6.2 Students are able to meditate for 15-20 minutes at a time

6.3 Students are able to do regular exercise and stretching, like yoga, qigong, tai chi, calisthenics, jogging, and other forms of physical fitness activities, like dance, sports, and/or martial arts

6.4 Students are able to develop healthy and balanced meal ideas based on bioregionally and homestead or school garden sourced foods

High School
(Identification, Comprehension, Application, Deep Analysis, Synthesis, Creation, Critique, Advanced Skill Acquisition, Community Building, & Service)

1. Comprehension, Analysis, & Critical Thinking
   1.1 Students are able to identify, define, apply, and synthesize permaculture concepts and practices as well as all permaculture ethics and principles and use them critically and creatively

   1.2 Students are able to problem solve using permaculture and explain why and how in a presentation, through teaching, social media, and other modalities

   1.3 Students are able to present and teach a variety of the skills and concepts to others using multimedia, public speaking, social media, and other modalities

   1.4 Students are able to use permaculture as a lens to problem solve in a diversity of novel and real-life scenarios ranging from ecosystemic to social, local to global

2. Knowledge
   2.1 Students are able to identify and map their bioregion and watershed digitally and on paper using topographic mapping, keyline geometry, and map and graphic editing programs

   2.2 Students are able to identify 50-100+ local native plants, 15-30+ native pollinators, 15-40+ local fungi, and 15-40+ local animals, their roles and interactions in the local ecosystem as well as any related indigenous history and uses

   2.3 Students are able to identify 100-200+ annual and perennial garden and food forest plants and their native origins and uses

   2.4 Students are able to identify, grow, harvest, process, preserve, and cook edible and medicinal mushrooms in both outdoor and indoor settings

   2.5 Students are able to identify, create detailed representations of, and teach the water cycle, the mineral cycle, the carbon cycle, and the global annual seasonal cycle in relation to the sun especially in relation to climate change and desertification

   2.6 Students are able to identify, describe, and present the different components of soil, the soil food web, photosynthesis in relation to the soil food web, and ways to improve soil and soil food web interactions

   2.7 Students are able to identify, describe, illustrate, and present the various interactions and function of trees and forests in relation ecosystems, natural cycles, precipitation, watersheds, soil, fungi, climate change, wildfires, and all biodiversity

3. Design
   3.1 Students are able to identify, map, and analyze their bioregion and watershed using keyline geometry and patterning

   3.2 Students are able to create an advanced permaculture design for a home site – one that addresses water, waste, food, soil building, energy, and shelter

3.3 Students are able to design, set up, and manage a small outdoor garden, indoor garden,
>   greenhouse garden, and larger outdoor garden

4. Regenerative Skills

4.1 Students are able to grow and teach others how to grow a broad diversity of plants from
>   seed, cutting, and tuber

4.2 Students are able to harvest and preserve a broad diversity of seed for long-term
>   storage, expanding and replenishing their own seed bank

4.3 Students are able to prepare and preserve food they have grown or helped grow in a
>   variety of ways, forming a diversity of meals and preserves that are shared, consumed
>   with the class, taken home, or sold

4.4 Students are able to make and apply thermophilic (hot) compost and vermicompost
>   (using worms) to process food waste as well as compost teas and extracts

4.5 Students are able to use a microscope to identify soil food web organisms and
>   determine soil, compost, compost extract, and compost tea quality

4.6 Students are able to cultivate fresh, salt, and brackish water aquatic plants and animals
>   in a small controlled environment as well as in a larger context

4.7 Students are able to design and install water harvesting and water management
>   earthworks and water storage systems

4.8 Students are able to filter water in a variety of ways, including sand and charcoal filters,
>   reedbeds and lagoons, mycoremediation, and phytoremediation to clean graywater
>   and blackwater

4.9 Students are able to calculate the volume of precipitation, evaporation,
>   moving water, and bodies of water

4.10 Students are able to design, install, and have experience repairing and maintaining a
>   diversity of earthworks, ponds, and dams

4.11 Students are able to design, build, and have experience repairing and maintaining
>   building structures using natural building methods

4.12 Students are able to identify and describe permaculture principles in business plans as
>   well as design, present, and critique regenerative business plans

4.13 Students are able to design renewable energy solutions on a home scale, and have
>   experience installing and maintaining /repairing renewable energy systems

4.14 Students have participated in regular acts of large-scale land restoration

4.15 Students have participated in regular acts of water and riparian restoration

5. Social & Self Skills

5.1 Students have participated in, planned, and helped manage many acts of service
>   connecting natural principles to social principles – providing a minimum of 30 hours of
>   community service

- **5.2** Students are able to identify, define, share, plan, promote, and participate in actions of compassion, empathy, and people care
- **5.3** Students are able to model nonviolent communication and restorative justice skills in everyday and conflict mediation settings

6. Self Smart Skills
- **6.1** Students are able to write out a life plan based on their holistic goals
- **6.2** Students are able to meditate for 20-50 minutes at a time
- **6.3** Students are able to maintain their physical health using exercise and stretching, like yoga, qigong, tai chi, calisthenics, jogging, and other forms of physical fitness activities, dance, sports, and/or martial arts
- **6.4** Students are able to develop healthy and balanced meal plans based on bioregionally and homestead or school garden sourced foods

Career Paths
College/University/Trade Schools/Internships

Since colleges, universities, internships, and trade schools have a wide range and broad diversity within them, I have not included standards for this level of education. Instead, these are general objectives and experiences related to a balanced professional educational experience that extends and builds upon the standards previously stated.

Learning Objectives & Experiences
- In-depth learning and hands-on, community-scale, and commercial-scale projects collaboratively planned, designed, and managed
- Immersion in professional contexts with experts
- Professional regenerative skill training and knowledge acquisition
- Professional community building skill training and event production
- Professional social & self smart skills

Current Standards Don't Lead to a Regenerative Future

Current standards are highly detailed and thoroughly iterative and repetitious. They attempt to cover every topic and subtopic possible deconstructing as far as is comprehensible into each topic area, so teachers can map out the entire year and still have standards and topics that require coverage – in other words, there's always sections of material that do not get covered. This leads to pacing guides made by districts to attempt to keep teachers and students on track for covering all the standards in a single academic year, but that creates an overall lack of depth in both teaching and student understanding – this makes sense to the outside observer: they ARE moving too fast, but it rarely occurs to those involved to stop this process and make fundamental changes to the system. The more we cram for tests and the more we focus on minute standards, the less we can get a deep analytical understanding of a topic, skill, or concept. Our current standards lack the fundamental human drive of meaning. Lacking principles and ethics, the standards have no strong ethical foundation or direction for human progress – you learn this set of information because that's how old you are and that's what you're taught at that age. The standards are uninspiring both to the student and the teacher. They do not focus on the student as a whole person, nor do they view education as a fluent holistic understanding of our world, but rather our current education standards focus on naming everything and memorizing as many names and their locations as possible – it is of the lowest two levels of cognition.

Today, you don't get far by being just like everyone else – that's the old system. Having the same answer and uniform performance is a factory education model, and in a world where automation, AI, and optimization are erasing jobs at an exponential rate (such that some are calling for a universal basic income), individuals with creative minds that can think critically are leading and innovating. College programs that once led to steady work, now lead to unemployment. It's more important today than ever to know how to support yourself – it's the job of K-12 education to prepare students for our world and economy. Our current standards are failing our students and leading them down a high stakes road to nowhere with no real life skills. If we are to prepare our students for both the future and the current economic, social, and environmental climate, they deserve and require the best tools, skills, and concepts to support them in this endeavor. These students are being asked to fix the neglect, damage, and degradation of thousands of years of human activity and to live in a new way that no one has fully adopted yet, and it is no doubt daunting to some degree, but the truth is they can be the heroes – this ball is in their hands but we have to prime and prepare them to run and make it across that finish line. We cannot do it without them – they are the future. We can put all the effort in but if they undo all our work when we age out, it was all for nothing. Our education system is the bridge to a stable, regenerative future if we can build it in time.

Another note on the separation of academic disciplines: if everything is separated out into parts of a whole and taught separately as if unrelated, how can we expect students to see how they interact, relate, rely upon each other, and form a complete whole system? After all, science is how we explain and make sense of our world and experiences, math is how we measure, calculate, and analyze aspects of those experiences, written and verbal communication and the arts are how we share our understanding, and history is the development of that understanding over time. Current schooling doesn't embrace this cross-curricular concept nor can they by following current standards – while we have humanities (history+English) lurking in some people's pasts, no one is taught holistically from a standards-based approach. This has to change if we want adults to think holistically and act on, rely upon, purchase, and vote for regenerative solutions to current threats to all global and local cultures, economies, and environments.

Many that see this book and don't pick it up will likely be of a specific "camp" — they'll be staked out and entrenched within a certain group, like NGSS, STEM, Environmental Literacy, or NSES. While each of these sets of standards has some validity and application, overall they are missing critical, fundamental components that would make their content engaging, empowering, regenerative, and authentic. They will undoubtedly all rely upon homework, worksheets, note taking, lecture-based instruction, and endless analysis.

Analysis paralysis makes sense when we look through the lens of K-12 education: you don't have to DO much of anything but be obedient and attentive, test well on verbal and math standards, and analyze endlessly using deconstructionism and critique (if you are lucky). Application, when it occurs, is inauthentic and does not build real-life skills, so the learning disconnects us from the natural world and the possibility of authentic learning itself. This leads to continued behavior of this sort as an adult, and we can see it especially in the adult environmental activist community and the environmental science standards: they can dissect everything that is wrong and perhaps even propose vague generalized plans for change, but they fail to make comprehensive change that goes viral in their own lives and thus their message is heard loud and clear: panic, be afraid, perhaps even be angry, but pass the message on, and wait for someone who knows what to do to take care of it all. It is a direct by-product of an education system that disempowers them and has them rely upon authority for answers and solutions. We are trained to analyze, to deconstruct, but not to act, to regenerate, and to steward wild ecosystems — that is what is lacking. While the current array of science standards have some merit, they lack understanding of how the world actually works: how science has evolved in the last 50 years and how to ethically manage our world and live our lives regeneratively.

Permaculture education standards don't necessarily displace any standards, but they do reinforce, clarify, support, inform, correct, extend, and demystify many standards in a variety of academic disciplines. The permaculture lens orders, prioritizes, and provides context for all standards, concepts, and education.

Next Generation Science Standards (NGSS)

NGSS prove that our kids are capable of a great deal, but is it helpful? Will all that information help them truly become citizen scientists, deflect climate disaster, and voluntarily adopt regenerative solutions in their daily lives?

To the Next Generation Science Standards, our world and universe is a Newtonian machine as is all life — that paradigm has been shattered by the science of the new century. The design of the NGSS standards is also disempowering, disconnecting, and focuses on technological, chemical, and engineered solutions with no focus on the intrinsic power of nature itself to heal, restore, and grow into an abundance when human activities are syntropic, partnering beneficially with natural cycles and patterns to perpetually generate new life. NGSS focuses on highly detailed but disconnected groupings of facts about and measurements of natural systems. NGSS is also often out of alignment - tasks are expected but no skills or solutions were part of the standards leading up to those expectations: NGSS asks students to help biodiversity after a sustained focus on evolution in HS-LS4 – misaligning the standards to outcome. Students cannot design from a numerical analysis perspective nor from evolutionary principles. Regenerative principles support biodiversity and overall ecosystemic health, not evolutionary principles which often focus on stress and pressure over time. They need to know how nature actually works and how they can interact with it. In addition, to evaluate designs without any principles of design or regenerative designs to source, with no testing, is unscientific. To research published data in mainstream media, selected readings, and standardized textbooks and to use numerical analysis of cost-benefit to evaluate which climate change plans are best is a convoluted exercise in disempowerment – examining and analyzing something you'll never be able to prove, disprove, or access the actual data sets. NGSS also focuses on computer simulations of what can go wrong – essentially perpetuated the doomsday predictions instead of focusing on an array of what is possible and what students can do. NGSS doesn't lead to a brighter future, but it does perpetuate the myth that analysis, technology, and authority will save us. NGSS doesn't provide solutions or regenerative skill building though it does expect students to generate or already have them. I include the NGSS standards, as they are a touchstone from the past allowing us to see where we've been and still are, and because of where we are going, it is absolutely necessary for many to orient themselves first. The NGSS standards do cover important topics, concepts, and ideas, but they are out of proportion, missing information, and organized poorly and without an understanding of current educational best practices and how students can incorporate and apply the standards into daily life. The fact is, all the standards available to educators have been up until now lacking the critical ingredients of the permaculture ethics, natural principles, personal growth, being a part of something larger than ourselves, and radical new insights in mycology, soil science, the water and carbon cycles, renewable energy, social science, educational psychology, and so much more. The reality is that science is NOT separate from our reality, that EVERYTHING is part of nature, and that WE are active participants in the natural world and scientists in our own lives, for better or worse. Allowing children and people of all ages to take upon themselves the responsibility for their own existence on this planet is incredibly important for them as human beings.

We need to know how to live responsibly, how to apply the scientific method ethically to all aspects of life, and how to live in a way that makes the world a better place through our daily lives. It can't be an empty gesture like the NGSS engineering design section; it has to be the main focus and the purpose of formal education.

While the NGSS (Next Generation Science Standards) are very thorough from an outdated perspective of science, they are strictly Newtonian, lack a true understanding of DNA and genetic inheritance, omit quantum physics altogether, and they lack a cohesive holistic connection to all other disciplines and all aspects of life, all of which tends to leave the learner feeling disconnected from the information. The detached nature of some academics finds its root in the detached nature of traditional education itself where in-depth learning leads to silos with their own secret language instead of deeper interconnections between disciplines. It turns the learner into a passive observer rather than an active participant. If we want active scientists, engaged regenerative stewards in local bioregions, and caring problem solvers, we need to train them in an empowering way: it must be active, engaging, and always linked to their world, micro to macro, in a meaningful, personal way.

There's a disempowerment of the learner when the focus begins with a look at the unseen and unverifiable world – the molecular and atomic ones. They are forced to take the teacher's word for it – the message is trust in authority instead of be curious and see for yourself. It is our job as educators to make natural cycles and scientific principles normal and understandable components of all aspects of daily life. Instead of starting with the molecular, it should start with the microscopic, in the verifiable, and allow them to do the actual work and see it for themselves. This relationship: micro to macro is key to building a strong scientific understanding of our natural world. Molecular, atomic, and quantum systems can be discussed, but the foundation should always be rooted in the verifiable and testing and application undertaken by the students themselves.

National Science Education Standards (NSES)

Similar to the NGSS offering, the National Science Education Standards lack ethics and a unifying foundation though they do have consistent unifying threads which are unquestioning faith in scientific authority, the scientific method, mathematical proofs, quantification, deconstructionism, reductionism, naming, and grouping. The NSES standards focus on creating the illusion of complete understanding and control over the natural world, which only deepens the disconnection between the students and the natural world and leads students and scientists to feel confident in outdated science and practice. Students need positive, beneficial interactions with real natural systems: micro to macro. NSES focuses, as NGSS does, initially on the molecular and atomic worlds – forcing students to learn based on their faith in their teachers, the textbooks, course materials, and their teacher's own faith in the topic instead of using scientific means to prove the concepts to the students or having the students prove the concepts themselves – this again is disempowering and creates learned helplessness and reliance on authority.

In classic science education fashion, NSES promotes the idea that scientists should be skeptical detectives or impartial observers instead of informed actors, caring, observant stewards, and passionate problem solvers working to serve people, the Earth, and the future holistically. The NSES standards continues the myth that humans and nature are separate. It does not clearly connect science to all other disciplines of academia and daily life. The NSES standards do not address climate change nor regenerative solutions.

Developing a Framework for Assessing Environmental Literacy (EL)

It must have been difficult creating the Environmental Literacy frameworks and assessments, given that the authors clearly are not environmentally literate and so have created an assessment that has no meaning because it is measuring something it cannot recognize or measure in a teaching and student population, as unacquainted with environmental literacy as they are. The NAAEE has been waiting for permaculture standards to provide the foundation for their frameworks and assessments. EL focuses on spreading awareness and having high expectations but without the scaffolding necessary for student success or accurate assessment of learning. NAAEE's focus on cognitive skills over hands-on, pragmatic, and professional skill building cheats students of the keys they'll need for regenerative success later on in life or currently in their home as well as robs the greater local community and even global collective of a myriad of holistic benefits.

It is clear the NAAEE Environmental Literacy assessment writers are very earnest in their desires to foment miraculous student breakthroughs through debate, study of accepted source materials (primarily mainstream media), and analysis of cost efficiency of different proposed technological solutions. It does not promote students sequestering carbon in massive amounts through student-led & designed bioregional projects restoring wildlife habitat, watersheds, riparian areas, wetlands, coastal regions, and more. These are simple things, but the NAAEE doesn't even approach them, nor does the NGSS.

Assessments of environmental literacy should be authentic, project-based, and regenerative. They should be bioregional and based on permaculture skills, strategies, principles, ethics, and standards. What the NAAEE and the NGSS need is a clear set of standards and fluency in natural systems and principles – that is what is found in this book and the companion textbook and workbook.

If we are to create a resilient, ethical, and regenerative culture, it begins with an education based on those concepts and focused on teaching those skills. It's time science education matures and becomes the lynchpin for all education through the permaculture lens.

IV. Lesson Plans

How to use this teacher's guide/lesson plans

Materials to be used: **The Permaculture Student 2 textbook and Workbook** set as well as *The Advanced Permaculture Student Online* course videos.

- Lecture/Reading
- PBL
- Group Learning
- Student-Centered
- Solutions-Based
- Authentic Application & Assessment
- Reflection

Direct instruction, reading, and note-taking still have a place in teaching, learning, and study. The best direct instruction is sometimes called storytelling or 1:1 coaching with an expert coach, but when students only get lectured, when it's boring, when it's tangential, when it's disorganized, etc., they disconnect and their learning suffers. Often lecture or reading can be shifted creatively by starting with a project or a problem they have to solve — this is a superpower of Project-Based Learning (PBL). Students get engaged and teachers simply provide materials and pathways for further study and research. They may provide a video of direct instruction, lecture, provide reading or audio materials, etc. Students discuss the projects with other students and the teacher(s) and can form teams or groups to do projects together, which is ideal. Solutions, mediums for expression, mediums of research, methods of research, and methods of expression can all differentiate to align to student learning modalities, interests, and prior knowledge.

Units 1 - 20 + Projects w/timelines and benchmarks for success

Note on NGSS: standards selected while in the same area of study as the standards often go well beyond, in a different direction entirely, or correct the current NGSS focus, which is fixated on the molecular and atomic world through the Newtonian lens only, which is flawed. Even the HS-LS2-7 standard to design solutions to reduce human interaction comes after a focus on mathematical analysis of ecosystems without any comprehension of how they work, function, are designed, or can be improved through human intervention. These standards disempower, focus on numerics instead of real skills and solutions, distract with their faux rigor (none of it embraces true complexity), and waste time when we need to teach the rising generation how to live regeneratively, partnering with nature syntropically.

Unit 1 - An Introduction to Permaculture & Pattern Literacy
Permaculture Education Standards: 1.1, 1.2, 1.3, 1.4
Next Generation Science Standards: HS-LS4-6, HS-ESS3-2

Objectives
- Students will be able to identify and define permaculture, the 3 Ethics, the principles of permaculture, pattern literacy, and edge effect.

Lesson Content
- Lecture/Presentation
- Reading/Listening
- Group Discussion
- Problem Solving using Permaculture in Novel Situations
- Nature Immersion & Observation of Natural Patterns

Lesson Activities
- Reading and discussing chapters 1-3 in *The Permaculture Student 2 Textbook*.
- Reading & Discussing pages 1-4 in *The Permaculture Student 2 Workbook*
- Watching & Discussing Week 1 of The Advanced Permaculture Student Online
- Begin to read and discuss *Braiding Sweetgrass* by Robin Wall Kimmerer as a class
- Begin to read and discuss *Climate, a New Story* by Charles Eisenstein as a class
- Visiting natural spaces and permaculture sites to observe natural patterns at play in the wild and in intentional settings.
- *What's Ethical?* - students can discuss and debate what is and what is not ethical in relation to the 3 Ethics: People Care, Earth Care, and Future Care.
- *Principles at Play* - students in groups form lists of examples related to each principle, seeking to bring a new example for each principle. Students afterward share, compare, and contrast their examples in group discussions.

- *Natural Pattern Sketches* - Using a pencil and paper in teams, go out and catalog as many of the different natural patterns that can be found in 15-20 minutes in a natural, garden, food forest, or wild space.
- *Snapshots* - students can use their phones or cameras to catalog and later present all the patterns their group found in a natural, garden, food forest, or wild space.

Assessments
- Group Discussion (observed and in direct participation)
- Verbal Quiz (individually or as a group)
- Identification & Comprehension Games/Exercises
- Written Quiz
- Essay, Video, Presentation, or Slideshow
- Student Presentations
- Non-graded Survey

Differentiations
Audiobook, adding background music, using songs, immersion in natural setting, taught while doing yoga, stretching, sketching, doodling, knitting, or something physical, tactile, and yet still automatic enough for them to listen, watch, and learn from the lesson content, more time, more discussion both 1:1 and/or in large and/or small group settings.

Unit 2 - Our Regenerative World

Permaculture Education Standards: 1.1, 1.2, 1.3, 1.4, 2.1, 2.2, 2.4, 2.5, 5.1
Next Generation Science Standards: HS-LS2-1, HS-LS2-2, HS-LS2-3, HS-LS2-4, HS-LS2-5, HS-LS2-6, HS-LS2-7, HS-LS2-8, HS-LS4-6, HS-ESS1-1, HS-ESS2-2, HS-ESS2-4, HS-ESS2-5, HS-ESS2-6, HS-ESS2-7, HS-ESS3-1, HS-ESS3-2, HS-ESS3-2, HS-ETS1-1, HS-ETS1-2, HS-ETS1-3, HS-ETS1-4

Objectives
- Students will be able to identify their climate as well as all other climates, their own bioregion, and the defining features of other bioregions
- Students will be able to understand, describe, and identify orographic, cyclonic, and convection effects in the landscape
- Students will be able to describe how and when dew, distillation, fog, frost, and condensation occur
- Students will be able to describe radiation in terms of solar radiation, color, albedo effect, and absorption
- Students will be able to describe and differentiate between convection, conduction, and radiation
- Students will be able to describe and identify a thermosiphon
- Students will be able to identify and describe thermal belts, different types of insulation and their efficacy, mitigating and amplifying factors for heat and heating, and how microclimates can be formed using these insights
- Students will be able to define, describe, and identify the behaviors of wind

- Students will be able to design and install a windbreak
- Students will be able to define, describe, and identify landscape effects such as continental, altitudinal, and latitudinal effects as well as atmospheric effects like precipitation, hurricanes, cyclones, and typhoons
- Students will be able to define, describe, identify, and create a graphic representation of the carbon cycle
- Students will be able to describe and identify how atmospheric CO_2 can be sequestered through photosynthesis and fungal partnership pathways into the soil, plants, and all biodiversity
- Students are able to describe and identify how atmospheric CO_2 levels relate to climate change and soil loss
- Students will be able to describe and use the Brittleness Scale

Lesson Content
- Lecture/Presentation/Video
- Reading/Listening/Watching
- Group Discussion
- Nature Immersion & Observation of Natural Systems
- Experiments
- Group Challenges & Activities
- Field Trips/Nature Visits

Lesson Activities
- Reading & Discussing chapter 4 in *The Permaculture Student 2*.
- Reading & Discussing pages 5-9 in *The Permaculture Student 2 Workbook*
- Watching & Discussing Week 2 of The Advanced Permaculture Student Online (excepting Trees & the Brittleness Scale)
- Continue to read and discuss Braiding Sweetgrass by Robin Wall Kimmerer as a class
- Continue to read and discuss *Climate, a New Story* by Charles Eisenstein as a class
- Watch NASA - A Year in the Life of Earth's CO_2 on Youtube & discuss the timing of the largest CO_2 releases as well as the seasonal sequestration of CO_2 in plant life.
- Visiting natural spaces and permaculture sites to observe natural systems and effects at work in the wild and leveraged and partnered with in intentional settings
- *Find your Climate & Climate Analogs* - Using the Köppen-Geiger Climate Classification system (on wikipedia.org), students will use the interactive map to find their spot on the map, related climate to that area, and the global climate analogs of that area
- *Build a Thermosiphon* - Students will use the sun to heat air and move a fan or turbine to show the properties of a thermosiphon in a group project challenge.
- *Take a Hike* - Students hike higher or lower in altitude, measuring temperature and elevation changes to demonstrate altitudinal effects
- *By Land or By Sea?* - Travel away or towards large bodies of water and coastal regions. Document how the landscape changes. This can be easily accomplished using Google

Earth or in coastal regions in extreme times of year (summer or winter) when the continental effect is most obvious
- *Solar Oven* - Students in groups will make solar ovens using cardboard, tape, and glass (or other resources as available). Using thermometers inside each, students can compete to see who can reach the highest temperatures longest
- *Keep the Heat* - Students in groups will strategize, plan, and build a heat insulating container for a quart mason jar of hot water -140°F/60°C or thereabouts, but all must be the same temperature at the beginning of the experiment. After an hour wait, the temperatures of the water inside the jars is tested. Students can compete to see who retains the most heat. A discussion about different types of insulation and their r-value can follow
- *Wind Mapping* - Using an anemometer, students can map the wind direction and speed on a site over the course of a day to see how wind direction and speeds change.
- *Carbon Cycle Mapping* - Students individually or in groups draw and illustrate the carbon cycle
- *Photosynthesis Mapping* - Students individually or in groups draw and illustrate the process of photosynthesis and its relationship to carbon sequestration in the soil and release of oxygen
- *Build a Windbreak* - Students will plan and plant a windbreak and measure the efficacy of the windbreak with an anemometer before and after.
- *Rain Shadow Search* - Using GoogleEarth or other satellite image maps to identify rain shadows individually, as a class, or in groups

Lesson Assessments
- Group Discussion (observed and in direct participation)
- Group Activities & Projects
- Verbal Quiz (individually or as a group)
- Identification Games/Exercises
- Written Quiz
- Student Presentations
- Non-graded Survey

Differentiations
Audiobook, adding background music, using songs, immersion in natural setting, taught while doing yoga, stretching, sketching, doodling, knitting, or something physical, tactile, and yet still automatic enough for them to listen, watch, and learn from the lesson content, more time, more discussion both 1:1 and/or in large and/or small group settings.

Unit 3 - Trees
Permaculture Education Standards: 1.1, 1.2, 1.3, 1.4, 2.2, 2.5, 2.7, 5.1
Next Generation Science Standards: HS-LS2-1, HS-LS2-2, HS-LS2-3, HS-LS2-4, HS-LS2-5, HS-LS2-6, HS-LS2-7, HS-LS2-8, HS-LS4-6, HS-ESS2-2, HS-ESS2-4, HS-ESS2-5, HS-ESS2-7, HS-ESS3-2, HS-ETS1-1, HS-ETS1-2, HS-ETS1-3, HS-ETS1-4

Objectives
- Students will be able to define, describe, and illustrate the functions and interactions of trees in natural settings
- Students will be able to identify interactions and functions of trees in actual natural settings
- Students will be able to describe wind effects on trees
- Students will be able to identify, define, and describe temperature effects on trees: evaporation (transpiration), condensation, and leaf color
- Students will be able to describe and identify trees relationship to precipitation, how trees can increase precipitation and interact with precipitation
- Students will be able to define and describe the conveyer belt of moisture concept

Lesson Content
- Lecture/Presentation/Video
- Reading/Listening/Watching
- Group Discussion
- Nature Immersion & Observation of Natural Systems
- Experiments
- Group Challenges & Activities
- Field Trips
- Community Service

Lesson Activities
- Reading & Discussing chapter 5 in *The Permaculture Student 2 Textbook*
- Finish Watching & Discussing Week 2 of The Advanced Permaculture Student Online
- Read and discuss as a class *The Hidden Life of Trees* by Peter Wohlleben
- Continue to read and discuss *Climate, a New Story* by Charles Eisenstein as a class
- Visiting a local forest, tree nursery, or permaculture orchard to observe, take notes, and identify natural processes in natural (or semi-natural) settings
- *Plant a Tree or Food Forest* - Students will plant a tree or more than one tree as part of larger forest or food system
- *Forest Biodiversity Mapping* - Students will map a section of forest, cataloging the plants and animals within the section, mapping out the locations of plants on a grid
- *Mapping a Trees Interactions* - Students individually or in groups make a visual (could be digital or made by hand) representing the interactions and functions of a tree in a natural forest. Students can present and describe the functions as groups as well.
- *Conveyer Belt of Moisture* - Students in groups or individually create visuals, animations, or models of the conveyor belt of moisture concept
- *Community Service* - Students can install food forests, windbreaks, and trees for those in need and in communal places with permission

Assessments
- Group Discussion (observed and in direct participation)
- Group Activities & Projects

- Verbal Quiz (individually or as a group)
- Identification Games/Exercises
- Written Quiz
- Student Presentations
- Non-graded Survey

Differentiations
Audiobook, adding background music, using songs, immersion in natural setting, taught while doing yoga, stretching, sketching, doodling, knitting, or something physical, tactile, and yet still automatic enough for them to listen, watch, and learn from the lesson content, more time, more discussion both 1:1 and/or in large and/or small group settings.

Week 4 - Water
Permaculture Education Standards: 1.1, 1.2, 1.3, 1.4, 2.5, 3.1, 4.7, 4.8, 4.9, 5.1
Next Generation Science Standards: HS-LS2-1, HS-LS2-2, HS-LS2-3, HS-LS2-4, HS-LS2-5, HS-LS2-6, HS-LS2-7, HS-LS2-8, HS-LS4-6, HS-ESS2-2, HS-ESS2-5, HS-ESS3-2, HS-ETS1-1, HS-ETS1-2, HS-ETS1-3, HS-ETS1-4

Objectives
- Students will be able to identify, define, illustrate, and present the water cycle
- Students will be able to identify, define, and illustrate watersheds in natural environments
- Students will be able to map water's path on a site and where it enters and leaves
- Students will be able to define and differentiate between types of cloud seeding, natural and manmade
- Students will be able to describe orographic and forest effects on the hydrological cycle
- Students will be able to define, design, utilize, and/or install water infiltration methods and earthworks
- Students will be able to define, design with, and strategically utilize soil, biological, landscape, and tank water storages
- Students will be able to define, design, install, and repair common dam/pond types and water harvesting earthworks
- Students will be able to describe historical and current water harvesting and storage techniques such as Petra's Nabatean water-harvesting systems and Brad Lancaster's homestead
- Students will be able to describe how natural ponds and springs form
- Students will be able to case a spring
- Students will be able to locate safe and suitable sites for a dam/pond
- Students will be able to define and describe the importance of the keyway/aquifuge/core of the dam wall
- Students will be able to calculate precipitation, runoff, and evaporation
- Students will be able to define, describe the importance of, design, and install spillways
- Students will be able to define, describe, and design including a trickle or spillway pipe

- Students will be able to define, describe, design, and install the backcut, side slopes, and freeboard of a dam/pond
- Students will be able to define, describe, design, and utilize rainwater storage, cleaning, and irrigation systems
- Students will be able to define, describe, design, and build swales as well as diversion banks and drains
- Students will be able to define, describe, and defend the ethics of earthworks and water
- Students will be able to define and describe common municipal graywater and blackwater treatment systems
- Students will be able to define, describe, design, and build home-scale graywater and blackwater remediation systems
- Students will be able to define, describe, and list water conservation methods, techniques, technologies, and strategies
- Students will be able to define, describe, design, build, and test water purification systems
- Students will be able to define, describe, design, build, utilize, and manage compost toilets
- Students will be able to define, describe, design, build, and manage natural swimming pools of drinking water quality and standard
- Students will be able to define, describe, and design with micro-hydropower

Lesson Content
- Lecture/Presentation/Video
- Reading/Listening/Watching
- Group Discussion
- Nature Immersion & Observation of Natural Systems
- Experiments
- Group Challenges & Activities
- Field Trip or Hike
- Community Service

Lesson Activities
- Reading & Discussing chapter 6 in _The Permaculture Student 2_.
- Reading & Discussing the Water Section in _The Permaculture Student 2 Workbook_
- Watching & Discussing Week 3 of The Advanced Permaculture Student Online
- Reading and discussing as a class and in small groups _Design and Construction of Small Earth Dams_ by K.D. Nelson.
- Reading and discussing as a class and in small groups Brad Lancaster's _Rainwater Harvesting for Drylands and Beyond (Vol. 2): Water-Harvesting Earthworks_
- Reading and discussing as a class and in small groups _Pond: Planning, Design, Construction_ by the NRCS
- Watching & Discussing Art Ludwig's _Laundry to Landscape Grey Water System_ DVD
- Read and discuss as a class and in small groups _Create an Oasis with Greywater_ and _Water Storage_ by Art Ludwig
- Continue to read and discuss _Braiding Sweetgrass_ by Robin Wall Kimmerer as a class

- Continue to read and discuss *Climate, a New Story* by Charles Eisenstein as a class
- Calculating Precipitation, Evaporation, Runoff, Stream/River Volume, & Pond/Lake Volume
- Touring the Local Municipal Sewage Treatment Plant
- *Hike Your Watershed* - Students will use maps to plot a trail to tour their watershed. Using GPS coordinates, students will verify their path and compare the elevations on paper to their GPS devices (cell phones and other WiFi devices)
- *Harvest Water* - Students as a group or in small teams in sections build a water harvesting swale or earthwork
- *Harvest Rainwater* - Students design and build a rainwater harvesting and cleaning system.
- *Community Service* - Students can design and install rainwater harvesting and cleaning systems for those in need or install water harvesting earthworks for those in need

Assessments
- Group Discussion (observed and in direct participation)
- Group Activities & Projects
- Verbal Quiz (individually or as a group)
- Identification Games/Exercises
- Written Quiz
- Essay (short or long)
- Student Presentations
- Non-graded Survey

Differentiations
Audiobook, adding background music, using songs, immersion in natural setting, taught while doing yoga, stretching, sketching, doodling, knitting, or something physical, tactile, and yet still automatic enough for them to listen, watch, and learn from the lesson content, more time, more discussion both 1:1 and/or in large and/or small group settings.

Unit 5 - Soils
Permaculture Education Standards: 1.1, 1.2, 1.3, 1.4, 2.5, 2.6, 3.2, 3.3, 4.4, 4.5, 5.1, 6.4
Next Generation Science Standards: HS-LS1-1, HS-LS1-3, HS-LS1-5, HS-LS2-1, HS-LS2-2, HS-LS2-3, HS-LS2-4, HS-LS2-5, HS-LS2-6, HS-LS2-7, HS-LS2-8, HS-LS4-6, HS-ESS2-6, HS-ESS2-7, HS-ESS3-2, HS-ETS1-1, HS-ETS1-2, HS-ETS1-3, HS-ETS1-4

Objectives
- Students will be able to define soil as well as describe how it is created in natural settings over time and to describe and demonstrate soil building through human action with mulch, composting, compost tea, compost extract, fermentation, biochar, and other methods.
- Students will be able to define and describe soil erosion
- Students will be able to define, describe, present, and illustrate the soil food web, its importance, and its functions

- Students will be able to use a microscope to identify members of the soil food web and differentiate between biologically safe (aerobic) and non-safe (anaerobic) compost, soils, and compost teas and extracts
- Students will be able to define and describe the importance and functions of fungi, bacteria, nematodes, protozoa, microarthropods, arthropods, worms, and small mammals and birds in relation to the soil food web
- Students will be able to define, describe, illustrate, and present Nitrogen in relation to plants, soils, pH, and types of Nitrogen
- Students will be able to define, describe, illustrate, and present the various processes of Nitrogen fixation
- Students will be able to define, describe, and critically think about plant roots and their growth habits
- Students will be able to define and describe macronutrients, their importance, and their various natural sources
- Students will be able to define, describe, and use pH in soil analysis and in relation to the Nitrogen cycle, fungi to bacteria ratios, vegetative and reproductive growth, annual and perennial plants, and plant succession
- Students will be able to define, describe, and use the Brittleness Scale in relation to soil analysis
- Students will be able to restore damaged and degraded soils using organic matter and soil biology
- Students will be able to explain why tilling soils destroys soil structure, releases carbon, changes pH, supports weed growth, and lowers nutritional densities of food
- Students will be able to define and describe soil water and gases
- Students will be able to describe how carbon sequestration in the soil works
- Students will be able to define and describe plant succession from bare ground to the old growth forests in relation to pH, fungi to bacteria ratios, and plant types as well as identify a bioregional succession example
- Students will be able to diagnose, troubleshoot, and treat problem soils, like compacted, hydrophobic, concreted, or cemented soils
- Students will be able to define, describe, and present ethical interactions with soil like no-till and low-tillage agriculture, perennial polycultures, aeration, subsoil plowing, conservation of soils, and building and enriching of soils in diverse ways
- Students will be able to build and maintain hugelkulturs and other soil-building earthworks
- Students will be able to use soil and earth as a building material
- Students will be able to define, describe, illustrate, and present how Nitrogen-fixing legumes interact with soil life
- Students will be able to use Nitrogen-fixing legumes to build and enrich soils
- Students will be able to use plants and fungi to remediate soils and define which plants are remediating what toxins, excess nutrients, or heavy metals
- Students will be able to use insects and worms to enrich or generate soil
- Students will be able to define, describe, create, and utilize inoculants of many kinds to enrich and build soils

- Students will be able to describe how windbreaks are soil-building mechanisms
- Students will be able to describe how ponds and ditches can build soils quickly through erosion and deposition
- Students will be able to define and describe how heirloom and landrace seeds provide a greater diversity of interactions in the soil food web in comparison to commercial F1 hybrids and GMOs
- Students will be able to define, describe, and present epigenetics, their inheritance mechanisms, and their importance
- Students will be able to describe and present how epigenetics are affected by everyday decisions, lifestyle choices, and habits, including what food they eat
- Students will be able to describe and present how GMOs are both unnecessary and a proven risk to natural gene pools
- Students will be able to use soil analyses to determine the current conditions of a landscape, some of its past management or mismanagement, and what remediating steps are needed
- Students will be able to decompose organic matter using both aerobic (thermophilic) composting and anaerobic fermentation
- Students will be able to define, describe, and make biochar, bokashi, DIY EM, and Korean natural farming indigenous microorganism preparations 1 through 5
- Students will be able to describe and provide examples of how plants accumulate nutrients in their growth and return them to the soil in their decomposition
- Students will be able to define, describe, and perform sheet mulching
- Students will be able to test and analyze soils using a variety of soil tests in terms of soil composition and the soil food web
- Students will be able to use a refractometer or BRIX meter to gauge overall plant and soil health

Lesson Content
- Lecture/Presentation/Video
- Reading/Listening/Watching
- Group Discussion
- Nature Immersion & Observation of Natural Systems
- Experiments
- Group Challenges & Activities
- Gardening & Soil Building Activities
- Community Service Activities

Lesson Activities
- Reading & Discussing chapter 7 in *The Permaculture Student 2*
- Reading & Discussing the Soil Section in *The Permaculture Student 2 Workbook*
- Watching & Discussing Week 4 & 5 of The Advanced Permaculture Student Online
- Watching the documentary *Symphony of the Soil* and discussing it as a class

- Read and discuss as a class and in small groups *Dirt: The Erosion of Civilizations* by David R. Montegomery
- Read and discuss as a class and in small groups *Adding Biology for Soil and Hydroponic Systems* by Dr. Elaine R. Ingham and Dr. Carole A. Rollins
- Read and discuss as a class and in small groups *10 Steps to Gardening with Nature* by Dr. Elaine R. Ingham and Dr. Carole A. Rollins
- Read and discuss as a class and in small groups *The Compost Tea Brewing Manual* by Dr. Elaine R. Ingham
- Read and discuss as a class and in small groups *Worms at Work* by Crystal Stevens
- Read and discuss as a class and in small groups *Teaming with Microbes* by Wayne Lewis and Jeff Lowenfels
- Read and discuss as a class and in small groups *Teaming with Nutrients* by Jeff Lowenfels
- Read and discuss as a class and in small groups *Earth Repair: A Grassroots Guide to Healing Toxic and Damaged Landscapes* by Leila Darwish
- Read and discuss as a class and in small groups *An Earth Saving Revolution* by Teruo Higa
- Read and discuss as a class and in small groups *Soil Not Oil* by Vandana Shiva
- Touring a local composting facility
- *Make Compost* - Students will make a variety of compost types in a variety of ways (vermicompost, thermophilic, & fermentation), analyze the results, and compare and contrast the methods
- *Make Compost Tea* - Students will be able to create a variety of compost teas and apply them to solve specific deficiencies or problems in the soil or plants
- *Make Compost Extract* - Students will make compost extract and apply it
- *Make Bokashi* - Students will make bokashi and apply it
- *Make DIY EM* - Students will make DIY EM and apply it
- *Make Biochar* - Students will make biochar and apply it
- *Soil Tests & Analysis* - Using a variety of soil tests, students will survey and test their soils at school, home, and in natural contexts within their bioregion
- *Soil Remediation* - Students will work in groups and make soil remediation plans and then put them into action – can include keyline subsoil plowing and soil ripping
- *Community Service* - Students will make compost, compost teas, and other beneficial soil amendments and donate their application or the quantity itself to those in need or to community spaces in need

Assessments
- Group Discussion (observed and in direct participation)
- Group Activities & Projects
- Verbal Quiz (individually or as a group)
- Identification Games/Exercises
- Performance Assessments
- Written Quiz
- Short & Long Essays
- Presentations & Dramatic Representations

- Student Presentations
- Non-graded Survey

Differentiations
Audiobook, adding background music, using songs, immersion in natural setting, taught while doing yoga, stretching, sketching, doodling, knitting, or something physical, tactile, and yet still automatic enough for them to listen, watch, and learn from the lesson content, more time, more discussion both 1:1 and/or in large and/or small group settings.

Unit 6 - Fungi

Permaculture Education Standards: 1.1, 1.2, 1.3, 1.4, 2.2, 2.4, 2.5, 2.6, 4.3, 5.1, 6.4
Next Generation Science Standards: HS-LS1-6, HS-LS2-1, HS-LS2-2, HS-LS2-3, HS-LS2-4, HS-LS2-5, HS-LS2-6, HS-LS2-7, HS-LS2-8, HS-LS4-6, HS-ESS1-6, HS-ESS2-2, HS-ESS2-4, HS-ESS2-6, HS-ESS2-7, HS-ESS3-2, HS-ETS1-1, HS-ETS1-2, HS-ETS1-3, HS-ETS1-4

Objectives
- Students will be able to define, describe fungi, its critical importance, its origins, its relationship to the origins of all life on Earth, and its current relationships to all life on Earth
- Students will be able to cultivate fungi in numerous ways indoors and outdoors, including liquid inoculation jars using airport lids
- Students will be able to identify the fungal phyla currently most well understood: ascomycota, basidiomycota, and glomeromycota as well as provide examples of members from each phyla
- Students will be able to describe and present fungi's role in the carbon cycle
- Students will be able to define, describe, cultivate, and apply mycorrhizal fungi
- Students will be able to define and differentiate between ectomycorrhizal and endomycorrhizal fungi
- Students will be able to take a spore print and assess it
- Students will be able to define, describe, and identify lichens
- Students will be able to use fungi to ferment food for eating purposes, like kefir, tempeh, etc.
- Students will be able to cultivate edible and medicinal fungi – especially mushrooms
- Students will be able to process mushrooms and mycelium to create a variety of fungal medicinal preparations like tinctures, teas and soups, and dried mycelium
- Students will be able to compost a variety of waste using fungi
- Students will be able to use fungi to remediate soils and to strengthen and heal plants and animals
- Students will be able to design, build, and utilize a sterile space to cultivate and work with fungi

Lesson Content
- Lecture/Presentation/Video
- Reading/Listening/Watching

- Group Discussion
- Nature Immersion & Observation of Natural Systems
- Experiments
- Group Challenges & Activities
- Foraging Hikes
- Spore Print Artwork & other Fungi and Mushroom Art
- Community Service

Lesson Activities
- Reading & Discussing chapter 8 in *The Permaculture Student 2*
- Reading & Discussing the Fungi Section in *The Permaculture Student 2 Workbook*
- Watching & Discussing Week 6 of The Advanced Permaculture Student Online
- Watching the documentary *Fantastic Fungi* and discussing it as a class
- Reading and discussing as a class and in small groups *Radical Mycology* by Peter McCoy
- Reading and discussing as a class and in small groups *Mycorrhizal Planet* by Michael Phillips
- *Oysters Are EASY* - Students will cultivate oyster mushrooms indoors or outdoors, possibly on spent coffee grinds and waste cardboard
- *Shiitake Logs* - Students will cultivate shiitake mushrooms on logs (ideally oak) in groups
- *Jar It Up* - Students will cultivate mycelium in liquid inoculation jars with airport lids
- *Let It Glow* - Students in teams will cultivate Panellus Stipticus, the bioluminescent fungi, in a liquid inoculation jar and then try to get it to fruit on a growing medium, which is a challenge, but the bioluminescence is visible in both the mycelial stage and the fruiting stage
- *King Stropharia Mushroom Bed* - Students will cultivate King Stropharia mushrooms in a garden style bed or raised garden
- *Let's Ferment!* - Students will make edible fermentations: water kefir, milk kefir, tempeh, kombucha, etc.
- *Dried Mycelium* - Students will strain and dry liquid culture mycelium
- *Mush Love* - Students in groups will prepare and serve a dish or meal featuring homegrown mushrooms
- *Going Foraging* - Students will forage for seasonal edible wild mushrooms with a focus on identification and mapping of location, not necessarily on eating or harvesting
- *Mycoremediation* - Students will use King Stropharia and Oyster inoculated straw and wood chips to filter water, testing the water going in and coming out, measuring to see how much of the contamination was removed.
- *Community Service* - using fungi to remediate an area, clean water, or to feed people, students will plan and perform a community service project with fungi

Assessments
- Group Discussion (observed and in direct participation)
- Group Activities & Projects
- Verbal Quiz (individually or as a group)
- Identification Games/Exercises

- Performance Assessment
- Written Quiz
- Student Presentations
- Non-graded Survey

Differentiations
Audiobook, adding background music, using songs, immersion in natural setting, taught while doing yoga, stretching, sketching, doodling, knitting, or something physical, tactile, and yet still automatic enough for them to listen, watch, and learn from the lesson content, more time, more discussion both 1:1 and/or in large and/or small group settings.

Unit 7 - Earthworks, Earth Resources, & Mapmaking
Permaculture Education Standards: 1.1, 1.2, 1.3, 1.4, 2.1, 3.2, 4.7, 4.10, 4.11, 4.14, 5.1, 5.2
Next Generation Science Standards: HS-LS4-6, HS-ESS3-2, HS-ETS1-1, HS-ETS1-2, HS-ETS1-3, HS-ETS1-4

Objectives
- Students will be able to define, describe, design, install, plant, and repair various types of earthworks
- Students will be able to define, describe, generate, analyze, and use topographic maps in design extensively
- Students will be able to define, describe, and demonstrate keyline geometry pattern understanding and mapping concepts
- Students will be able to use a topographic map to find the path of water on a site and trace the watershed
- Students will be able to plot a line on contour using an A-frame
- Students will be able to build and calibrate an A-frame
- Students will be able to calculate slope and use slope to make design decisions
- Students will be able to select ideal building locations
- Students will be able to define, describe, and use various levels
- Students will be able to use natural building materials in construction of various building projects large and small
- Students will be able to use animals, hand tools, and powered machines to create earthworks

Lesson Content
- Lecture/Presentation/Video
- Reading/Listening/Watching
- Group Discussion
- Nature Immersion & Observation of Natural Systems
- Experiments
- Group Challenges & Activities

Lesson Activities
- Reading & Discussing chapter 9 in *The Permaculture Student 2*
- Reading & Discussing the Earthworks Section in *The Permaculture Student 2 Workbook*
- Watching & Discussing Week 7 of The Advanced Permaculture Student Online
- Begin reading and discussing as a class and in small groups Darren Doherty's *Regrarian's Handbook*
- Begin reading and discussing as a class and in small groups PA Yeomans' *The Keyline Plan* and *The Challenge of Landscape*
- *Make a Topographic Map* - Students will create a topographic map using a variety of methods including GoogleEarth, SketchUp, and more
- *Build & Use an A-frame* - Students will build and properly use A-frames to plot a line on contour
- *Design & Install Earthworks* - Students will design earthworks to solve a problem or serve a beneficial purpose and install them
- *Build an Earth-Sheltered Greenhouse* - Students will design an earth-sheltered greenhouse or walipini and install it
- *Community Service* - Students will plan, design, and install earthworks for a communal or public space or for someone or those in need or face a difficulty that earthworks could address

Assessments
- Group Discussion (observed and in direct participation)
- Group Activities & Projects
- Verbal Quiz (individually or as a group)
- Identification Games/Exercises
- Written Quiz
- Mapmaking
- Performance Assessment
- Student Presentations
- Non-graded Survey

Differentiations
Audiobook, adding background music, using songs, immersion in natural setting, taught while doing yoga, stretching, sketching, doodling, knitting, or something physical, tactile, and yet still automatic enough for them to listen, watch, and learn from the lesson content, more time, more discussion both 1:1 and/or in large and/or small group settings.

Unit 8 - Permaculture Processes & Frameworks
Permaculture Education Standards: 1.1, 1.2, 1.3, 1.4, 2.1, 2.2, 2.3, 3.1, 3.2, 3.3, 4.7, 4.8, 4.9, 4.10, 4.12, 4.14, 4.15, 5.1, 5.2, 6.1
Next Generation Science Standards: HS-LS4-6, HS-ESS3-2, HS-ETS1-1, HS-ETS1-2, HS-ETS1-3, HS-ETS1-4

Objectives
- Students will be able to define, describe, and apply permaculture concepts and themes in design
- Students will be able to define, describe, and utilize a diversity of permaculture processes and frameworks
- Students will be able to analyze resources that they may be able to make ethical decisions with those resources
- Students will be able to apply the scientific method through the lens of the 3 Ethics
- Students will be able to analyze, discuss, and present all elements in a system in terms of their variety, products, needs, and behaviors
- Students will be able to create functional designs that demonstrate extended observation, study, and reflection
- Students will be able to define, describe, apply, and present zone planning and analyses in the design process
- Students will be able to define, describe, and create flow diagrams
- Students will be able to define and describe energy flow on a given site
- Students will be able to determine and calculate orientation, slope, and aspect
- Students will be able to calculate sun angle and design with annual sun angle changes in mind
- Students will be able to define, describe, and design for the soundscape
- Students will be able to define, describe, and utilize timing, incremental design, and strategies that create and extend yields in design
- Students will be able to use define and describe holism
- Students will be able to define, describe, and demonstrate holistic goal setting
- Students will be able to define, describe, and utilize the holistic management framework
- Students will be able to define, describe, and apply the four principles of holistic management in design and management plans
- Students will be able to define, describe, and apply the Keyline Scale of Permanence and The Regrarians Platform
- Students will be able to define, describe, and apply keyline patterning in design
- Students will be able to define, describe, illustrate, locate, and map the keypoint and keyline of a primary land unit
- Students will be able to apply keyline patterning and the Regrarians Platform to develop business plans and projected ROI and EROEI for each plan
- Students will be able to adopt rewilding and wildlife and native ecosystem restoration into their designs

Lesson Content
- Lecture/Presentation/Video
- Reading/Listening/Watching
- Group Discussion
- Nature Immersion & Observation of Natural Systems

- Experiments
- Group Challenges & Activities
- Mapmaking

Lesson Activities
- Reading & Discussing chapter 10 in *The Permaculture Student 2*
- Watching & Discussing Week 8 of The Advanced Permaculture Student Online
- Continue reading and discussing as a class and in small groups Darren Doherty's *Regrarian's Handbook*
- Continue reading and discussing as a class and in small groups PA Yeomans' *The Keyline Plan* and *The Challenge of Landscape*
- Watch and discuss as a class *Polyfaces,* the documentary on Joel Salatin's family farm
- *Map the Keypoint & Keyline* - Students in groups using pencil and paper topographic maps or GoogleEarth will locate the keypoint and keyline in a diversity of landscapes and map them on Google Earth and/or on paper by hand
- *Design a Site with The Regrarians Platform* - Students will design a site using the Regrarians Platform (RP) in groups and as a class
- *Design an Orchard with RP* - Students will design an orchard using the Regrarians Platform (RP) in groups and as a class on paper with pencil as well as digitally on GoogleEarth
- *Design a Ranch with RP* - Students will design a ranch for a holistic management operation using the Regrarians Platform (RP) in groups and as a class
- *Analyze an Element* - Students will analyze, discuss, and present elements in a system in terms of their variety, products, needs, and behaviors
- *Install a Keypoint Dam or a Keyline Swale* - Students will analyze a site, locate the keypoint and keyline, and design and install a keypoint dam and/or keypoint swale or diversion drain
- *Rewilding* - Students will rewild or restore a wilderness area to a wild, regenerative, autonomous ecosystemic state
- *Holistic Goal Setting* - Students will set holistic goals for themselves
- *Holistic Planning* - Students will plan a business, farm, ranch, or service using the holistic management framework and principles
- *Community Service* - Students will plan, design, and perform design services and installations for those in need and for communal areas with consent

Assessments
- Group Discussion (observed and in direct participation)
- Group Activities & Projects
- Verbal Quiz (individually or as a group)
- Identification Games/Exercises
- Performance Assessment
- Written Quiz
- Regrarians Platform & Essay Rationale
- Holistic Goal Setting & Planning
- Student Presentations

- Non-graded Survey

Differentiations

Audiobook, adding background music, using songs, immersion in natural setting, taught while doing yoga, stretching, sketching, doodling, knitting, or something physical, tactile, and yet still automatic enough for them to listen, watch, and learn from the lesson content, more time, more discussion both 1:1 and/or in large and/or small group settings.

Unit 9 - Food Forests & Gardens

Permaculture Education Standards: 1.1, 1.2, 1.3, 1.4, 2.1, 2.2, 2.3, 2.4, 3.2, 3.3, 4.1, 4.2, 4.3, 4.4, 4.5, 5.1, 5.2, 6.4
Next Generation Science Standards: HS-LS3-1, HS-LS3-2, HS-LS3-3, HS-LS4-6, HS-ESS3-2, HS-ETS1-1, HS-ETS1-2, HS-ETS1-3, HS-ETS1-4

Objectives
- Students will be able to grow, transplant, harvest, and care for annual and perennial plants from seed to seed
- Students will be able to grow plants from tuber and cutting
- Students will be able to save seed and begin their own seed bank
- Students will be able to identify 100+ local native plants and their indigenous history and uses
- Students will be able to identify 250+ annual and perennial garden and food forest plants
- Students will be able to grow, harvest, and cook outdoor mushrooms
- Students will be able to prepare fresh food from a food forest or garden into a meal
- Students will be able to define, identify, describe, design, and install beneficial plant guilds
- Students will be able to identify, define, and describe the succession of a forest from bare ground to old growth forest
- Students will be able to identify, define, describe, and grow bioregional plant successions
- Students will be able to identify, define, describe, design, and install all the layers of the forest
- Students will be able to demonstrate a diversity of planting, harvesting, and managing techniques for gardens and food forests
- Students will be able to use multiple plant guides to analyze plants and their characteristics
- Students will be able to utilize beneficial fungi in the food forest and garden, including growing edible mushrooms
- Students will be able to define, describe, present, and cultivate polycultures
- Students will be able to describe the difference between monocultures and polycultures, especially in terms of soil and plant health
- Students will be able to describe the difference between annuals and perennials, especially in terms of soil disturbance and carbon sequestration
- Students will be able to design, design, and manage gardens in a variety of regenerative ways
- Students will be able to define, describe, and design crop rotation plans

- Students will be able to define, describe, and utilize regenerative biointensive gardening methods
- Students will be able to define, describe, design, and create beneficial insect and pollinator habitat
- Students will be able to define, describe, design, and create wild bird habitat
- Students will be able to graft fruit tree branches
- Students will be able to prune and train trees and other plants
- Students will be able to define, describe, present, and practice food forest and garden management techniques that are regenerative
- Students will be able to define, describe, present, and practice hand pollination of plants and practice plant breeding

Lesson Content
- Lecture/Presentation/Video
- Reading/Listening/Watching
- Group Discussion
- Nature Immersion & Observation of Natural Systems
- Experiments
- Group Challenges & Activities
- Gardening & Landscaping

Lesson Activities
- Reading & Discussing chapter 11 in *The Permaculture Student 2*
- Reading & Discussing the Trees and Food Forests and Gardening sections in *The Permaculture Student 2 The Workbook*
- Watching & Discussing Week 9 & 10 of The Advanced Permaculture Student Online
- Continue reading and discussing as a class and in small groups *Braiding Sweetgrass* by Robin Wall Kimmerer
- Reading and discussing as a class and in small groups *The Resilient Gardener* and *Breed Your Own Vegetable Varieties* by Carol Deppe
- Reading and discussing as a class and in small groups *The Market Gardener* by Jean Martin Fortier
- Reading and discussing as a class and in small groups *The Urban Farmer* by Curtis Stone
- Reading and discussing as a class and in small groups *Understanding Roots* by Robert Kourik
- Reading and discussing as a class and in small groups *The Holistic Orchard* by Michael Phillips
- Reading and discussing as a class and in small groups *Perennial Vegetables* by Eric Toensmeier
- Reading and discussing as a class and in small groups *Integrated Forest Gardening: The Complete Guide to Polycultures and Plant Guilds in Permaculture Systems*
- Reading and discussing as a class and in small groups *Farming the Woods* by Ken Mudge and Steve Gabriel

- Referencing & reading selections from Seed Savers Exchange's *The Seed Garden*
- Referencing & reading selections from *Trees for Gardens, Orchards, and Permaculture*
- *Plant a Garden* - Students will plan, design, install, and manage a garden from seed
- *Plant a Food Forest* - Students will plan, design, install, and manage a food forest
- *Save the Seed* - Students will build and maintain the school's heirloom, non-GMO seed bank and their own personal heirloom, non-GMO seed collections
- *Harvest Feast* - Students will prepare meals with the garden and food forest harvests, serving it to each other and/or community members
- *Preserve the Harvest* - Students will preserve food in a variety of ways
- *Breed Your Own* - Students will hand pollinate to try to breed a new variety from the F1 hybrid that follows the next season — this can take 3-10 years to stabilize the new variety, but it is also an easy project. Squash crosses, for instance, are easy and yield quick results.
- *Pure Seed Presentations* - Students in groups will create multimedia presentations on why heirloom seeds, seed saving, and bioregional seed savers are vital
- *An All Season Tree* - Students will graft cuttings from early, mid, and late season varieties onto one tree to create an "all season" tree
- *Community Seed Swap* - Students will host a community seed swap using seeds from their gardens or the school's gardens to share with the community and each other
- *Share the Seed* - Students will research, network, and send seeds to those in need of heirloom, non-GMO seeds, for instance, other bioregional and climate analog schools and communities starting their garden programs
- *Share the Harvest* - Students will research, network, plan, and share the fresh and preserved abundance of food and seed with community members in need
- *Community Service* - Students will plan, design, and install gardens and food forests for the community or those in need

Assessments
- Group Discussion (observed and in direct participation)
- Group Activities & Projects
- Verbal Quiz (individually or as a group)
- Identification Games/Exercises
- Written Quiz
- Performance Assessment
- Long and Short Essays
- Student Presentations
- Non-graded Survey

Differentiations
Audiobook, adding background music, using songs, immersion in natural setting, taught while doing yoga, stretching, sketching, doodling, knitting, or something physical, tactile, and yet still automatic enough for them to listen, watch, and learn from the lesson content, more time, more discussion both 1:1 and/or in large and/or small group settings.

Unit 10 - Climate Specifics & Climate Lists

Permaculture Education Standards: 1.1, 1.2, 1.3, 1.4, 2.1, 3.1, 3.2, 3.3
Next Generation Science Standards: HS-LS4-6, HS-ESS3-2, HS-ETS1-1, HS-ETS1-2, HS-ETS1-3, HS-ETS1-4

Objectives
- Students will be able to define, describe, and design gardens, food forests, earthworks, aquaculture, animal systems, energy systems, water harvesting and storage, soil, and housing in tropical climates
- Students will be able to define, describe, and design gardens, food forests, earthworks, aquaculture, animal systems, energy systems, water harvesting and storage, soil, and housing in temperate climates
- Students will be able to define, describe, and design gardens, food forests, earthworks, aquaculture, animal systems, energy systems, water harvesting and storage, soil, and housing in arid climates
- Students will be able to define, describe, and utilize a diversity of climate-specific techniques and strategies for all broad climate types
- Students will be able to describe and define climate specific plants and fungi food forest layers as well as garden plants and field crops

Lesson Content
- Lecture/Presentation/Video
- Reading/Listening/Watching
- Group Discussion
- Nature Immersion & Observation of Natural Systems
- Experiments
- Group Challenges & Activities

Lesson Activities
- Reading & Discussing chapter 12-14 in _The Permaculture Student 2_
- Watching & Discussing Week 9 & 10 of _The Advanced Permaculture Student Online_
- Continue reading and discussing as a class and in small groups _Braiding Sweetgrass_ by Robin Wall Kimmerer
- Watching & Discussing Dr. Willie Smits's TEDtalk _How to Restore a Rainforest_
- Watching & Discussing Sustainable Design Masterclass' _Terraforming the Desert with Neal Spackman_
- Watching & Discussing _Polyfaces,_ the documentary
- Watching & Discussing _Inhabit,_ the documentary
- Watching & Discussing _Greening the Desert with Geoff Lawton,_ the documentary
- Watching & Discussing _Green Gold,_ the John D. Liu documentary
- _Design By Climate_ - Students in groups will design and present similar sites in temperate, tropical, and arid climates to highlight the similarities and differences between them —

farming through three climates, ranching through climates, and food forestry through three climates
- *Site Designs* - Students will create both current and future "dream" site designs and present them

Assessments
- Group Discussion (observed and in direct participation)
- Group Activities & Projects
- Verbal Quiz (individually or as a group)
- Identification Games/Exercises
- Performance Assessments
- Written Quiz
- Student Presentations
- Non-graded Survey

Differentiations
Audiobook, adding background music, using songs, immersion in natural setting, taught while doing yoga, stretching, sketching, doodling, knitting, or something physical, tactile, and yet still automatic enough for them to listen, watch, and learn from the lesson content, more time, more discussion both 1:1 and/or in large and/or small group settings.

Unit 11 - Aquaculture
Permaculture Education Standards: 1.1, 1.2, 1.3, 1.4, 4.6, 4.10, 4.15, 5.1, 5.2
Next Generation Science Standards: HS-LS4-6, HS-ESS3-2, HS-ETS1-1, HS-ETS1-2, HS-ETS1-3, HS-ETS1-4

Objectives
- Students will be able to define, describe, and design regenerative aquaculture
- Students will be able to design and manage regenerative indoor and outdoor aquaculture systems
- Students will be able to define, describe, and present the aquaculture's plant and trophic layers
- Students will be able to define, describe, design, present, setup, and manage an aquaculture food chain
- Students will be able to define, describe, and calculate yields, feed rates, and stocking rates
- Students will be able to define, describe, illustrate, and present the aquatic Nitrogen cycle
- Students will be able to define and describe the importance and roles pH, temperature, and salinity play in aquatic systems
- Students will be able to design systems to maintain and manage aquaculture system's oxygen levels
- Students will be able to design, build, and manage islands, decks, and floating gardens
- Students will be able to design and manage rice systems
- Students will be able to define, design, build, and manage natural swimming pools

- Students will be able to define and design chinampas, channels, and canals
- Students will be able to rewild and restore a diversity of riparian, wetland, and coastal areas in fresh, salt, and brackish waters
- Students will be able to define, describe, design, build, and manage large-scale ocean, sea, and lake restoration and rewilding
- Students will be able to define, describe, and present the keys to ocean restoration
- Students will be able to define and describe vertical ocean farming, artificial reefs, and coastal microclimates
- Students will be able to describe and present how the oceans are vital to all life on land

Lesson Content
- Lecture/Presentation/Video
- Reading/Listening/Watching
- Group Discussion
- Nature Immersion & Observation of Natural Systems
- Experiments
- Group Challenges & Activities
- Community Service

Lesson Activities
- Reading & Discussing chapter 15 in <u>The Permaculture Student 2</u>
- Watching & Discussing Week 11 of *The Advanced Permaculture Student Online*
- Reading and Discussing as a class and in small groups <u>Let Water Do The Work</u> by Bill Zeedyk
- Watch & Discuss Dan Barber's TEDtalk *How I Fell In Love With A Fish*
- *Setup an Aquaponics System* - Students will design and build an aquaponics system with fish and plants
- *Setup an Aquaculture System* - Students will design and build an aquaculture system with fish and plants
- *Setup a Pond* - Students will design, add, and manage plants, fish, amphibians, and animals in an outdoor pond site
- *Natural Swimming Pool* - Students will design and build a natural swimming pool
- *Cultivate Spirulina or Duckweed* - Students will design and create a spirulina or duckweed cultivation and harvesting system
- *Rewild a Wetland* - Students will rewild a wetland with native species and a variety of restoration methods
- *Rewild the Coast* - Students will rewild a coastal area with native species and a variety of restoration methods
- *Ocean Farming* - Students will grow kelp, mussels, or some other mollusk or seaweed using
- *CleanUp the Coast & Riparian Areas* - Students will clean up a coastal or riparian area by picking up garbage and removing impediments or dangers to wildlife
- *Community Service* - Students will plan and take action to restore or rewild a riparian area or build a pond in a communal space or for those in need

Assessments
- Group Discussion (observed and in direct participation)
- Group Activities & Projects
- Verbal Quiz (individually or as a group)
- Identification Games/Exercises
- Written Quiz
- Performance Assessment
- Student Presentations
- Non-graded Survey

Differentiations
Audiobook, adding background music, using songs, immersion in natural setting, taught while doing yoga, stretching, sketching, doodling, knitting, or something physical, tactile, and yet still automatic enough for them to listen, watch, and learn from the lesson content, more time, more discussion both 1:1 and/or in large and/or small group settings.

Unit 12 - Renewable Energy
Permaculture Education Standards: 1.1, 1.2, 1.3, 1.4, 3.2, 4.13, 5.1, 5.2
Next Generation Science Standards: HS-LS4-6, HS-ESS3-2, , HS-ESS3-4, HS-ETS1-1, HS-ETS1-2, HS-ETS1-3, HS-ETS1-4

Objectives
- Students will be able to define, describe, design, install, and manage a diverse range of renewable energy systems
- Students will be able to define, describe, and use a variety of hand tools
- Students will be able to define, describe, and calculate Energy Return on Energy Invested (ERORI)
- Students will be able to generate energy using bioregional resources in ethical regenerative ways
- Students will be able to analyze and critique energy generation and storage systems
- Students will be able to analyze, critique, and adapt energy systems for longevity and the least amount of energy and cost input
- Students will be able to define, describe, and utilize multiple ways that energy can be stored
- Students will be able to define, describe, and utilize a diversity of passive energy, heating, and cooling systems
- Students will be able to define, describe, and utilize gasification, biofuel, biogas, and other natural combustible energy sources
- Students will be able to define, describe, design, illustrate, and present carbon capture systems for combustion released CO_2

- Students will be able to define, describe, and create pneumatic systems using water and air pressure
- Students will be able to define, describe, and create systems that rely upon the sun for a large variety of energy-saving purposes

Lesson Content
- Lecture/Presentation/Video
- Reading/Listening/Watching
- Group Discussion
- Nature Immersion & Observation of Natural Systems
- Experiments
- Group Challenges & Activities
- Community Service

Lesson Activities
- Reading & Discussing chapter 16 in *The Permaculture Student 2*
- Reading & Discussing the *Alternative Energy* section in *The Permaculture Student 2 The Workbook*
- Watching & Discussing Week 12 of *The Advanced Permaculture Student Online*
- Reading & Discussing as a class *Just Enough: Lessons in Living Green from Traditional Japan* by Azby Brown
- Reading and Discussing as a class and in small groups *DIY Solar Power* by Micah Toll
- *Conservation Comes First* - Students in groups will brainstorm and research energy conservation tips and methods to use less electricity, water, heat, cooling, gas, and fuel.
- *Build a Mini-Trompe* - Students in groups will use running water to create a miniature trompe and run a pneumatic turbine or turn an anemometer. Students groups can compete to see which design produces the most energy or potential energy
- *Student Expert Groups* - Students will form groups and each focus on a different renewable energy to critically analyze and rate. They will present their analyses and ratings to the students (and community if possible)
- *Go Off-Grid* - Students will in groups or together as a class plan, design, and install a renewable energy solution for a building, service, or system
- *Community Service* - Students will plan, design, and install a renewable energy solution in a community space or for those in need

Assessments
- Group Discussion (observed and in direct participation)
- Group Activities & Projects
- Verbal Quiz (individually or as a group)
- Identification Games/Exercises
- Performance Assessment
- Experiments
- Written Quiz

- Long and Short Essays
- Student Presentations
- Non-graded Survey

Differentiations

Audiobook, adding background music, using songs, immersion in natural setting, taught while doing yoga, stretching, sketching, doodling, knitting, or something physical, tactile, and yet still automatic enough for them to listen, watch, and learn from the lesson content, more time, more discussion both 1:1 and/or in large and/or small group settings.

Unit 13 - Urban Permaculture

Permaculture Education Standards: 1.1, 1.2, 1.3, 1.4, 2.4, 2.6, 2.7, 3.2, 3.3, 4.4, 4.6, 4.8, 4.11, 4.13, 4.16, 5.1, 5.2
Next Generation Science Standards: HS-LS4-6, HS-ESS3-2, , HS-ESS3-4, HS-ETS1-1, HS-ETS1-2, HS-ETS1-3, HS-ETS1-4

Objectives
- Students will be able to define, design, adapt, and retrofit regenerative urban systems, techniques, methods, and strategies
- Students will be able to design systems for as well as teach and practice urban black and grey water purification methods
- Students will be able to design for small spaces, for low-light, for no natural light, and for indoor systems as well as build and manage them
- Students will be able to grow food in containers and in raised garden beds indoors and outdoors
- Students will be able to design, describe, illustrate, build, and manage urban energy generation systems
- Students will be able to will be able to define, describe, design, and utilize regenerative forms of transportation
- Students will be able to define, describe, and demonstrate ancient and indigenous traditions for living regeneratively in urban or heavily populated spaces
- Students will be able to define, describe, and list current urban regenerative systems, explain how they are regenerative, and critique how they could be improved
- Students will be able to define, describe, and practice rooftop farming

Lesson Content
- Lecture/Presentation/Video
- Reading/Listening/Watching
- Group Discussion
- Nature Immersion & Observation of Natural Systems
- Experiments
- Group Challenges & Activities
- Performance Assessments

Lesson Activities
- Reading & Discussing chapter 17 in *The Permaculture Student 2*
- Watching & Discussing Week 14 of *The Advanced Permaculture Student Online*
- Reading & Discussing as a class *The Permaculture City* by Toby Hemenway
- Reading & Discussing as a class *The Forest City* by PA Yeomans
- Reading & Discussing as a class *Retrosurbia* by David Holmgren
- *Retrofit New York City (or any city)* - Students in groups will create plans and designs to regeneratively retrofit New York City or another large city and present them to the class (and possibly community)
- *Design a Permaculture City* - Students in groups will create plans and designs for a permaculture city and present them to the class (and possibly community)
- *Design a Permaculture City Block or Neighborhood* - Students in groups will create plans and designs for a permaculture city block or neighborhood and present them to the class (and possibly community)
- *Regenerate an Apartment* - Students in groups or independently will create plans and designs for a permaculture apartment and balcony and present them to the class (and possibly community)
- *Community Service* - Students in groups or as a class will plan, design, and manage an event or social media presenting urban permacultural solutions to the general public or the community

Assessments
- Group Discussion (observed and in direct participation)
- Group Activities & Projects
- Verbal Quiz (individually or as a group)
- Identification Games/Exercises
- Performance Assessment
- Written Quiz
- Long and Short Essays
- Student Presentations
- Non-graded Survey

Differentiations
Audiobook, adding background music, using songs, immersion in natural setting, taught while doing yoga, stretching, sketching, doodling, knitting, or something physical, tactile, and yet still automatic enough for them to listen, watch, and learn from the lesson content, more time, more discussion both 1:1 and/or in large and/or small group settings.

Unit 14 - Permatecture

Permaculture Education Standards: 1.1, 1.2, 1.3, 1.4, 2.5, 2.6, 2.7, 3.2, 3.3, 4.8, 4.10, 4.11, 4.12, 4.13, 5.2
Next Generation Science Standards: HS-LS4-6, HS-ESS3-2, , HS-ESS3-4, HS-ETS1-1, HS-ETS1-2, HS-ETS1-3, HS-ETS1-4

Objectives
- Students will be able to define and describe permatecture
- Students will be able to define, describe, demonstrate in design, and present the permatecture principles
- Students will be able to define, describe, and build with a variety of natural building materials
- Students will be able to define, describe, list, design with, and utilize bioregional regenerative building material sources
- Students will be able to demonstrate, explain, and present the application of the 3 Ethics of permaculture to building materials, building construction, worker wages, accessibility, repairability, and all aspects dealing with the construction
- Students will be able to efficiently stack functions in architecture and design, showing multiple functions for one space or element

Lesson Content
- Lecture/Presentation/Video
- Reading/Listening/Watching
- Group Discussion
- Nature Immersion & Observation of Natural Systems
- Experiments
- Group Challenges & Activities

Lesson Activities
- Reading & Discussing chapter 18 in The Permaculture Student 2
- Watching & Discussing Week 15 of The Advanced Permaculture Student Online
- Reading & Discussing the Permatecture Section in The Permaculture Student 2 The Workbook
- Reading & Discussing as a class The $50 & Up Underground House Book and The Earth-Sheltered Solar Greenhouse Book by Mike Oehler
- Reading & Discussing as a class The Forest Garden Greenhouse by Jerome Osentowski
- Reading & Discussing as a class Earth-Sheltered Houses by Rob Roy
- Reading & Discussing as a class The Hand-Sculpted House by Ianto Evans, Michael G. Smith, and Linda Smiley
- Reading & Discussing as a class Water Storage: Tanks Cisterns, Aquifers, and Ponds by Art Ludwig
- The Permatecture Home - Students will independently and/or in groups design and present a permatecture home and permatectural construction plan

- *Permatecture Team Build* - Students will as a class design, plan the construction of, and build a permatecture building following permatecture principles in construction and design
- *Community Service* - Students will as a class design, plan the construction of, and build a permatecture building for a communal space, service, or for those in need following permatecture principles in construction and design

Assessments
- Group Discussion (observed and in direct participation)
- Group Activities & Projects
- Verbal Quiz (individually or as a group)
- Identification Games/Exercises
- Student Presentations
- Written Quiz
- Non-graded Survey

Differentiations
Audiobook, adding background music, using songs, immersion in natural setting, taught while doing yoga, stretching, sketching, doodling, knitting, or something physical, tactile, and yet still automatic enough for them to listen, watch, and learn from the lesson content, more time, more discussion both 1:1 and/or in large and/or small group settings.

Unit 15 - Social Permaculture
Permaculture Education Standards: 1.1, 1.2, 1.3, 1.4, 5.1, 5.2, 5.3, 5.3, 6.1, 6.2, 6.3, 6.4
Next Generation Science Standards: HS-LS4-6, HS-ETS1-1, HS-ETS1-2, HS-ETS1-3, HS-ETS1-4

Objectives
- Students will be able to define, describe, identify, design, apply, and demonstrate social permaculture
- Students will be able to define, describe, identify, apply, and demonstrate principles for people systems
- Students will be able to define, describe, identify, apply, and demonstrate people care including self care
- Students will be able to define, describe, identify, apply, and present the concepts of care of future, decentralization, and autonomy in people systems
- Students will be able to define, describe, present, and identify community patterns, behaviors, structures, and components
- Students will be able to define, describe, present, design, and identify intentional communities, land trusts, and villages
- Students will be able to define, describe, and present the ethical imperative to empower women as well as the environmental and socioeconomic benefits of empowering women
- Students will be able to define, describe, present, and identify the functions and behaviors of governance

- Students will be able to define, describe, present, identify, and demonstrate sociocracy in action
- Students will be able to define, describe, present, identify, and demonstrate holacracy in action
- Students will be able to define, describe, present, identify, and demonstrate nonviolent communication (NVC) as well as NVC's components, parts, and process
- Students will be able to define, describe, present, identify, and demonstrate restorative justice and restorative circles

Lesson Content
- Lecture/Presentation/Video
- Reading/Listening/Watching
- Group Discussion
- Nature Immersion & Observation of Natural Systems
- Experiments
- Group Challenges & Activities

Lesson Activities
- Reading & Discussing chapter 19 in *The Permaculture Student 2*
- Watching & Discussing Week 16 of *The Advanced Permaculture Student Online*
- Reading & Discussing the Social Permaculture section in *The Permaculture Student 2 The Workbook*
- Reading & Discussing as a class *The More Beautiful World Our Hearts Know Is Possible* by Charles Eisenstein
- Continue Reading and Discussing as a class *Braiding Sweetgrass* by Robin Wall Kimmerer
- Reading & Discussing as a class *Holistic Management* 3rd Edition by Alan Savory
- Reading & Discussing as a class *The Motivation Manifesto* and *High Performance Habits* by Brendon Burchard
- Reading & Discussing as a class *Nonviolent Communication: A Language of Life* by Marhsall B. Rosenberg, Ph.D
- Reading & Discussing as a class *Nonviolent Communication: Companion Workbook* by Lucy Leu
- Reading & Discussing as a class *We the People: Consenting to a Deeper Democracy* by John Buck and Sharon Villines
- Reading & Discussing as a class *Holacracy: the New Management System for a Rapidly Changing World* by Brian J. Robertson
- Continue Reading & Discussing as a class *Retrosuburbia* by David Holmgren
- *Meditation Challenge* - Students will set a daily meditation goal of 20 minutes or more first thing in the morning for 30 days or more. Following the challenge, they will present on their experiences with meditation
- *Gratitude Journal* - Students will keep a gratitude journal for 30 days writing down 3-5 specific examples from their day they are grateful for in the last 15-20 minutes before sleep each night, and they will present on their experiences through journaling with the class

- *Practicing NVC* - Students will use role playing games to re-enact examples found NVC suggested reading
- *Practicing People Care* - Students in groups or independently will research, present, plan, and demonstrate a diversity of examples of people care
- *Practicing Self Care* - Students in groups or independently will research, present, and demonstrate a diversity of examples of self care including healthy diet, regular physical exercise, meditation, spirituality, life learning, journaling, high performance habits, and yoga – could also be organized into an event of presentations and demonstrations or workshops
- *Practicing Restorative Circles* - Students will use role-playing games to re-enact examples found NVC suggested reading
- *People Care Essays* - Students will write essays on how people care has manifested throughout history, how people care can be applied today, and how people care is important for the future
- *Build Community* - Students will plan, design, and manage a community building event with people care as the focus
- *Community Service* - Students in groups or as a class will plan, design, and manage a community service event focused on individuals in need or educating the public on a common need related to people care

Assessments
- Group Discussion (observed and in direct participation)
- Group Activities & Projects
- Verbal Quiz (individually or as a group)
- Identification Games/Exercises
- Written Quiz
- Long and Short Essays
- Presentations
- Performance Assessments
- Non-graded Survey

Differentiations
Audiobook, adding background music, using songs, immersion in natural setting, taught while doing yoga, stretching, sketching, doodling, knitting, or something physical, tactile, and yet still automatic enough for them to listen, watch, and learn from the lesson content, more time, more discussion both 1:1 and/or in large and/or small group settings.

Unit 16 - The Regenerative Economy
Permaculture Education Standards: 1.1, 1.2, 1.3, 1.4, 2.1, 2.2, 2.3, 2.4, 2.5, 2.6, 2.7, 3.1, 3.2, 3.3, 4.1, 4.2, 4.3, 4.4, 4.5, 4.6, 4.7, 4.8, 4.9, 4.10, 4.11, 4.12, 4.13, 4.14, 4.15, 4.16, 5.1, 5.2, 6.1
Next Generation Science Standards: HS-LS4-6, HS-ESS3-2, HS-ETS1-1, HS-ETS1-2, HS-ETS1-3, HS-ETS1-4

Objectives
- Students will be able to define, describe, and present on the regenerative economy
- Students will be able to define, describe, present on, and participate in the gift economy
- Students will be able to define, describe, present, and demonstrate on reciprocity
- Students will be able to define, describe, and present on the third industrial revolution
- Students will be able to define, describe, design, and present on the regenerative companies
- Students will be able to define, describe, and present on the difference between syntropy and entropy
- Students will be able to define, describe, and present on the bioregional economy
- Students will be able to define, describe, and present on the alternative and local currencies
- Students will be able to define, describe, and present on micro-loans, micro-insurance, and micro-franchises
- Students will be able to define, describe, and present on the difference between the formal and informal economies of cultures and communities
- Students will be able to define, describe, design, and present on non-profit organizations
- Students will be able to define, describe, and present on cooperatives/co-ops/buying clubs
- Students will be able to define, describe, design campaigns for, and present on crowdfunding
- Students will be able to define, describe, design, and present on CSAs and farm-share farms
- Students will be able to define, describe, design, and present on stacked careers and products as well as business guilds
- Students will be able to define, describe, design, and present regenerative business plans
- Students will be able to define, describe, and present leading examples of regenerative businesses and non-profit organizations
- Students will be able to define, describe, and present on the fibershed concept

Lesson Content
- Lecture/Presentation/Video
- Reading/Listening/Watching
- Group Discussion
- Nature Immersion & Observation of Natural Systems
- Experiments
- Presentations
- Performance Assessments
- Group Challenges & Activities

Lesson Activities
- Finish Reading & Discussing chapter 19 in *The Permaculture Student 2*
- Watching & Discussing Week 17 of *The Advanced Permaculture Student Online*
- Reading & Discussing as a class the Economic subsection of the Social Permaculture section in *The Permaculture Student 2 The Workbook*

- Reading & Discussing as a class *The Regenerative Business* by Carol Sanford
- Continue Reading & Discussing as a class *Dark Emu* by Bruce Pascoe
- Continue Reading & Discussing *Holistic Management* 3rd Edition by Alan Savory
- Continue Reading & Discussing as a class *Braiding Sweetgrass* by Robin Wall Kimmerer
- Continue Reading & Discussing as a class *The More Beautiful World Our Hearts Know Is Possible* by Charles Eisenstein
- Reading & Discussing as a class *The Regenerative Organic Certification Standards* by the Rodale Institute
- *Your Regenerative Business Plan* - Student in groups or independently will research, design, present, and, if possible, demonstrate a regenerative business plan
- *Regenerative Researchers* - Students in groups or independently will research regenerative businesses and then present on them to each other
- *Regenerative Dream Jobs* - Students will research and present regenerative careers and jobs
- *Business Planning with the Regrarians Platform* - Students independently or in groups will create a business plan using the Regrarians Platform
- *Business Planning with Holistic Management* - Students independently or in groups will create a business plan using holistic management methods
- *Regenerative Career Fair* - Students will plan, design, and host a regenerative career fair with talks, regenerative job offers, regenerative business showcases, and educational materials
- *Community Service* - Students will create community service and non-profit business plans, so they can be adopted by anyone in the community or put into action by the students

Assessments
- Group Discussion (observed and in direct participation)
- Group Activities & Projects
- Verbal Quiz (individually or as a group)
- Identification Games/Exercises
- Business Planning
- Performance Assessments
- Written Quiz
- Non-graded Survey

Differentiations
Audiobook, adding background music, using songs, immersion in natural setting, taught while doing yoga, stretching, sketching, doodling, knitting, or something physical, tactile, and yet still automatic enough for them to listen, watch, and learn from the lesson content, more time, more discussion both 1:1 and/or in large and/or small group settings.

Unit 17 - Regenerative Agriculture, Ranching, & Homesteading

Permaculture Education Standards: 1.1, 1.2, 1.3, 1.4, 2.1, 2.2, 2.3, 2.4, 2.5, 2.6, 2.7, 3.1, 3.2, 3.3, 4.1, 4.2, 4.3, 4.4, 4.5, 4.6, 4.7, 4.8, 4.9, 4.10, 4.11, 4.12, 4.13, 5.1, 5.2
Next Generation Science Standards: HS-LS4-6, HS-ESS3-2, HS-ESS3-3, HS-ETS1-1, HS-ETS1-2, HS-ETS1-3, HS-ETS1-4

Objectives
- Students will be able to define, describe, identify, and practice regenerative agriculture, ranching, and homesteading
- Students will be able to define, describe, identify, and practice biological farming, carbon farming, perennial farming, natural farming, and holistic management of cattle.
- Students will be able to define, describe, identify, and present a diversity of regenerative agricultural, ranching, and homesteading examples across climates and contexts
- Students will be able to analyze a piece of property and suggest what form of regenerative agriculture would be suitable, it's potential ROI, and why

Lesson Content
- Lecture/Presentation/Video
- Reading/Listening/Watching
- Group Discussion
- Nature Immersion & Observation of Natural Systems
- Experiments
- Presentations
- Group Challenges & Activities
- Performance Assessment

Lesson Activities
- Reading & Discussing chapter 20 in _The Permaculture Student 2_
- Watching & Discussing Week 18 of _The Advanced Permaculture Student Online_
- Watching & Discussing _The Salatin Semester_ by Verge Permaculture and Acres USA
- Reading & Discussing as a class the _Regenerative Ranching_ and _Regenerative Agriculture_ sections in _The Permaculture Student 2 The Workbook_
- Continue Reading & Discussing as a class _Dark Emu_ by Bruce Pascoe
- Reading & Discussing as a class _The Biggest Estate on Earth: How Aborigines Made Australia_ by Bill Gammage
- Reading & Discussing as a class _Call of the Reed Warbler_ by Charles Massy
- Continue Reading & Discussing as a class _Holistic Management_ 3rd Edition by Alan Savory
- Continue Reading & Discussing as a class _The Regrarians Handbook_ by Darren Doherty
- Reading & Discussing as a class _The Carbon Farming Solution_ by Eric Toensmeier
- Reading & Discussing as a class _The Market Gardener_ by Jean-Martin Fortier
- Reading & Discussing as a class _The Urban Farmer_ by Curtis Stone
- Reading & Discussing as a class _The Lean Farm_ by Ben Hartman

- *Your Regenerative Farm, Ranch, or Homestead* - Student in groups or independently will research, design, present, and, if possible, demonstrate a regenerative farm, ranch, or homestead plan
- *Intern/Apprentice/Volunteer* - Students will serve as an intern, apprentice, or volunteer at least 5 full working days at more than one bioregional regenerative site, business, or service
- *Community Service* - Students in groups or as a class will plan, design, and manage a community service event focused on regenerative agriculture, ranching, and homesteading

Assessments
- Group Discussion (observed and in direct participation)
- Group Activities & Projects
- Verbal Quiz (individually or as a group)
- Identification Games/Exercises
- Written Quiz
- Long and Short Essays
- Performance Assessments
- Non-graded Survey

Differentiations
Audiobook, adding background music, using songs, immersion in natural setting, taught while doing yoga, stretching, sketching, doodling, knitting, or something physical, tactile, and yet still automatic enough for them to listen, watch, and learn from the lesson content, more time, more discussion both 1:1 and/or in large and/or small group settings.

Unit 18 - Permaculture in Action (Working Examples)
Permaculture Education Standards: 1.1, 1.2, 1.3, 1.4, 2.1, 2.2, 2.3, 2.4, 2.5, 2.6, 2.7, 3.1, 3.2, 3.3, 4.1, 4.2, 4.3, 4.4, 4.5, 4.6, 4.7, 4.8, 4.9, 4.10, 4.11, 4.12, 4.13, 4.14, 4.15, 4.16, 5.1, 5.2, 6.1
Next Generation Science Standards: HS-LS4-6, HS-ESS3-2, HS-ESS3-3, HS-ETS1-1, HS-ETS1-2, HS-ETS1-3, HS-ETS1-4

Objectives
- Students will be able to define, describe, present, list examples of, and demonstrate permaculture in action, from small-scale to large-scale
- Students will be able to define, describe, present, list examples of, design, and participate in large-scale land restoration projects
- Students will be able to define, describe, present, list examples of, design, and participate in large-scale ocean restoration projects
- Students will be able to define, describe, present, list examples of, design, and participate in large-scale riparian restoration projects
- Students will be able to define, describe, present, list examples of, design, and participate in large-scale watershed restoration projects
- Students will be able to define, describe, and present on the Loess Plateau Restoration project

- Students will be able to define, describe, and present on the Al Baydha project
- Students will be able to define, describe, and present on the regenerative cultural examples, histories, food systems, methods, and techniques of the Aboriginal Australians, the Ancient Aztecs (the Triple Alliance), Native Americans, and all the Indigenous peoples of the world
- Students will be able to define, describe, design, and create a legacy project for their class or community, a beneficial long-lasting addition to the community
- Students will be able to define, define, and present holistic goals, holistic life plans, well developed regenerative business/non-profit plans, and restoration and rewilding plans of all kinds, including social, political, economic, environmental, and more.

Lesson Content
- Lecture/Presentation/Video
- Reading/Listening/Watching
- Group Discussion
- Nature Immersion & Observation of Natural Systems
- Experiments
- Group Challenges & Activities

Lesson Activities
- Reading & Discussing chapter 21 in *The Permaculture Student 2*
- Watching & Discussing Week 19 of *The Advanced Permaculture Student Online*
- Reading & Discussing as a class the *Water & Ocean Restoration* and *Land Restoration* sections in *The Permaculture Student 2 The Workbook*
- Continue Reading & Discussing *Braiding Sweetgrass* by Robin Wall Kimmerer
- Continue Reading & Discussing as a class *The Biggest Estate on Earth: How Aborigines Made Australia* by Bill Gammage
- Continue Reading & Discussing as a class *Call of the Reed Warbler* by Charles Massy
- Continue Reading & Discussing as a class *Holistic Management* 3rd Edition by Alan Savory
- Continue Reading & Discussing as a class *The Regrarians Handbook* by Darren Doherty
- Watching & Discussing *Lessons of the Loess Plateau* by John D. Liu
- Watching & Discussing *A Real History of Aboriginal Australians - Bruce Pascoe TEDtalk*
- Watching & Discussing *Taking Root: The Vision of Wangari Maathai*
- Watching & Discussing *Growing Power - A Model for Urban Agriculture*
- Watching & Discussing *This Country Isn't Just Carbon Neutral – It's Carbon Negative | Tshering Tobgay* (TEDtalk)
- Watching & Discussing Dr. Willie Smits's TEDtalk, *How to Restore a Rainforest*
- Watching & Discussing Sustainable Design Masterclass' *Terraforming the Desert with Neal Spackman*
- Watching & Discussing *Polyfaces,* the documentary
- Watching & Discussing *Inhabit,* the documentary
- Watching & Discussing *Greening the Desert with Geoff Lawton,* the documentary
- Watching & Discussing *Green Gold,* the John D. Liu documentary

- *Take Action!* - Students individually and in groups will design with and practice permaculture in a diversity of situations ranging from small-scale to large-scale, simple to complex, and in a diversity of climates and contexts, urban to suburban to rural to remote
- *Large-Scale Restoration* - Students as a class will plan, design, and put into action a large-scale restoration project if possible – if not, students will take part in a large-scale restoration project: ocean, riparian, land, and watershed restoration
- *Regeneration Through Time* - Students will be able to define, describe, identify, present, and, if possible, demonstrate indigenous regenerative stories, concepts, techniques, and methods as well as how indigenous knowledge has informed or can inform present-day models
- *Permaculture Awareness Events* - Students in groups or as a class will plan, design, and manage a community service event focused on showcasing permaculture in action and providing workshops for the community – students will teach workshops
- *Community Service* - Students in groups or as a class will plan, design, and manage a community service using permaculture

Assessments
- Group Discussion (observed and in direct participation)
- Group Activities & Projects
- Verbal Quiz (individually or as a group)
- Identification Games/Exercises
- Written Quiz
- Non-graded Survey

Differentiations
Audiobook, adding background music, using songs, immersion in natural setting, taught while doing yoga, stretching, sketching, doodling, knitting, or something physical, tactile, and yet still automatic enough for them to listen, watch, and learn from the lesson content, more time, more discussion both 1:1 and/or in large and/or small group settings.

Unit 19 - The Designer's Mindset
Permaculture Education Standards: 1.1, 1.2, 1.3, 1.4
Next Generation Science Standards: HS-LS4-6, HS-ESS3-2, HS-ESS3-3, HS-ETS1-1, HS-ETS1-2, HS-ETS1-3, HS-ETS1-4

Objectives
- Students will be able to define, describe, identify, and demonstrate the designer's mindset
- Students will be able to reflect, write, and reflect on all that they've learned, experienced, and designed during the course
- Students will present their reflections to the class and community

Lesson Content

- Lecture/Presentation/Video
- Reading/Listening/Watching
- Group Discussion

Lesson Activities
- Reading & Discussing chapter 22 in *The Permaculture Student 2*
- Watching & Discussing Week 20 of *The Advanced Permaculture Student Online*
- Reading & Discussing as a class the *Your Next Step: The Future* section in *The Permaculture Student 2 The Workbook*
- Continue Reading & Discussing *Braiding Sweetgrass* by Robin Wall Kimmerer
- Student Reflection Presentations
- Reflection Essays

Assessments
- Group Discussion (observed and in direct participation)
- Non-graded Survey
- Long or Short Essay

Differentiations
Audiobook, adding background music, using songs, immersion in natural setting, taught while doing yoga, stretching, sketching, doodling, knitting, or something physical, tactile, and yet still automatic enough for them to listen, watch, and learn from the lesson content, more time, more discussion both 1:1 and/or in large and/or small group settings.

Unit 20 - Permaculture Design Certification

Permaculture Education Standards: 1.1, 1.2, 1.3, 1.4, 2.1, 2.2, 2.3, 2.4, 2.5, 2.6, 2.7, 3.1, 3.2, 3.3, 4.7, 4.9, 4.10
Next Generation Science Standards: HS-LS4-6, HS-ESS3-2, HS-ESS3-3, HS-ETS1-1, HS-ETS1-2, HS-ETS1-3, HS-ETS1-4

Objectives
- Students will be able to be able to demonstrate their learning with a final design for a PDC certification (which is 72 hrs of education + a design demonstrating learning)
- Students will share and present the designs with other students and/or the larger community to get feedback and share permacultural solutions

Lesson Content
- Lecture/Presentation/Video
- Reading/Listening/Watching
- Group Discussion
- Presentations

Lesson Activities

- Reading, Discussing, & Filling Out *The Permaculture Student 1 The Workbook*
- Students will use the Regrarians Platform to design a site and create a map
- Students will create sophisticated and professional site maps using SketchUp, GoogleEarth, CAD, GIS, Adobe Creative Suite, other mapping and graphic software, and/or pencil, paper, and rolling protractor rulers
- Students will use present their site designs to their classmates and community

Assessments
- Group Discussion (observed and in direct participation)
- Performance Assessment
- Site Designs & Rationales
- Non-graded Survey
- Long or Short Essay
- Presentations
- Performance Assessment

Differentiations
More time, different mediums of expression than presentation in front of the group, multiple options for presentation medium and design type and context

Unit 21 - Advanced Permaculture Design Certification

Standards: 1.1, 1.2, 1.3, 1.4, 2.1, 2.2, 2.3, 2.4, 2.5, 2.6, 2.7, 3.1, 3.2, 3.3, 4.7, 4.8, 4.9, 4.10, 4.11, 4.12, 4.13, 4.14, 4.15, 4.16, 6.1
Next Generation Science Standards: HS-LS4-6, HS-ESS3-2, HS-ESS3-3, HS-ETS1-1, HS-ETS1-2, HS-ETS1-3, HS-ETS1-4

Objectives
- Students will be able to be able to demonstrate their learning with a final design for a PDC certification (which is 72 hrs of education + a design demonstrating learning)
- Students will share and present the designs with other students and/or the larger community to get feedback and share permacultural solutions

Lesson Content
- Lecture/Presentation/Video
- Reading/Listening/Watching
- Group Discussion
- Performance Assessment
- Self Assessment
- Benchmarks

Lesson Activities
- Reading, Discussing, & Utilizing *The Permaculture Student 2 The Workbook* to improve student designs and to launch their designs into reality

- Students will use the Regrarians Platform to design a site, business, farm, ranch, or another type of project and write out the 3-5 steps, or benchmarks, that must be completed to make this project a reality
- Students will create sophisticated and professional site maps using SketchUp, GoogleEarth, CAD, GIS, Adobe Creative Suite, other mapping and graphic software, and/or pencil, paper, and rolling protractor rulers
- Students will use project proposals to their classmates and community for feedback and greater insight
- *Create Your Benchmark Timeline* - Students will take their 3-5 steps, their Benchmarks, and set them on a timeline: how long will each step realistically take? What are the sub-steps for each step? What will be the benchmark deadlines?
- *Start Your Journey* - Students will begin working on their First Benchmark
- *Benchmark Reports* - Students will report in to the community, teacher, and other students when benchmarks are prematurely achieved, going to be late, actually achieved, or are not possible – in this situation, new benchmarks and possibly a new project proposal are needed

Assessments
- Group Discussion (observed and in direct participation)
- Performance Assessment
- Site Designs & Rationales
- Non-graded Survey
- Benchmark Reflections
- Presentations
- Performance Assessment

Differentiations
More time, different mediums of expression than presentation in front of the group, multiple options for presentation medium, project type, and design type and context

Unit 22 - APDC Reflection & Certification Awarding
Standards: 1.1, 1.2, 1.3, 1.4, 2.1, 2.2, 2.3, 2.4, 2.5, 2.6, 2.7, 3.1, 3.2, 3.3, 4.1, 4.2, 4.3, 4.4, 4.5, 4.6, 4.7, 4.8, 4.9, 4.10, 4.11, 4.12, 4.13, 4.14, 4.15, 4.16, 5.1, 5.2, 6.1, 6.4, HS-ETS1-1, HS-ETS1-2, HS-ETS1-3, HS-ETS1-4

Objectives
- Students will be able to be able to demonstrate their learning with an authentic project for an advanced permaculture design certification (which is 145+ hrs of education + an authentic project based on student-generated designs demonstrating learning)
- Students will share and present the project benchmarks and final reflection with other students and/or the larger community to get feedback and share advanced permacultural solutions

Lesson Content
- Lecture/Presentation/Video
- Reading/Listening/Watching
- Group Discussion
- Performance Assessment
- Self Assessment
- Presentations

Lesson Activities
- Students will complete their project benchmarks, fill out the reflective surveys, and present their reflections and experience with their fellow students and the community
- *Reflect On Your Journey* - Students will reflect on their benchmarks, their successes, their failures, what they would have done differently, and what their plans are for the future
- *Share Your Journey* - Students will present their reflections on their benchmarks, their successes, their failures, what they would have done differently, and what their plans are for the future. This can be in-person in front of an audience to the students or community, online in a class group, filmed, at a community event, written, an audio recording, or even an animation or presentation set to an audio recording

Assessments
- Group Discussion (observed and in direct participation)
- Performance Assessment
- Site Designs & Rationales
- Non-graded Survey
- Long or Short Essay
- Presentations
- Performance Assessment

Differentiations
More time, multiple mediums of expression, multiple platforms and locations for expression

Advanced Permaculture Design Certification (APDC) Project Proposal Samples

Guiding Questions
- *What's your project and its goals?*
- *How is your project regenerative holistically?*
- *What is your timeline?*
- *What will be your benchmarks for success?*

Things to Remember
- **These Are Self-Imposed** - *You set the bar & the checkpoints!*
- **While It Seems Simple** - *Especially when Steps & Benchmarks are back-to-back, but start planning in that progression and it'll begin to flow. You'll discover that each step and benchmark likely have their internal set of steps.*
- **We Are Here To Help** - *Ask Matt, other teachers, and the class for advice & feedback.*
- **There Is No Time Minimum or Limit** - *You could take a year or longer to finish your project and reach all the goals you set, but how much more the sweeter your certification that you earned by setting the bar that high.*
- **Get Feedback & then Launch** - *Getting feedback not just from Matt but from other teachers and fellow students is immensely powerful and helpful. It can save you time, money, and energy — it can even protect you if you miss something vital!*
- **Launching Means Sharing Your Proposal** - *This can be written, video, or audio. It can be in the comments, using Google Docs, or even in our Facebook Group.*
- **Share Your Benchmarks** - *They are designed to share with the class and everyone you know, so you can be cheered by your peers and get critical feedback to overcome inevitable obstacles.*
- **Certification Is Earned** - *When benchmarks are fulfilled, the completed project or business launch is reflected upon, and your final thoughts are shared with the class inside the course, in the Facebook group, or through releasing the communication through Matt. This final sharing can be via writing, video, or audio.*

Sample #1
"Fire-Proofing" the California Sierras with Polycultures, Earthworks, Native Plants, and Patchwork Burns

Level of Difficulty: EXTREME
Level of Benefit: MASSIVE

Goal
To prevent destructive wildfires, protect wildlife and human habitation, and to reestablish the natural pyrophytic cycles of low burning seasonal fire.

How Is It Regenerative?
This project provides a simple but radical model and call to action for all pyrophytic landscapes to change their relationship with fire and become proactive regenerators of their bioregional landscapes – healing forests, biodiversity, and watersheds while protecting people and the planet.

Timeline
2-5 years, bioregionally

Benchmarks
1. **Silviculture Program Established** - Forests are being thinned, remove branches up to 6m, managed regeneratively, and are being grown naturally in place from seed
2. **Strategic Fire Breaks Established** - Using earthworks, water harvesting, some logging, & controlled burns in large corridors designed to stymie large forest fires before reaching major highways, cities, and towns are in place
3. **Watershed Preservations Established** - Using keyline design, dam removal, conservation, water harvesting and spreading earthworks, native biodiversity, watersheds are restored, large destructive dams are removed, water is spread and infiltrated, and everything is designed to promote and restore native biodiversity in a thriving autonomous system
4. **Patchwork Burn Systems Established** - Using Keyline Geometry and Wind Mapping, a system for burning safely is established bioregionally in which burn districts are created for citizens and government organizations to manage properly the timing and practice of seasonal low burning

Sample #2
An Advanced Permaculture Homestead
Level of Difficulty: HARD
Level of Benefit: LARGE

Goal
Create a demonstration site of an advanced permaculture homestead: regeneratively sourcing necessary basics like food, power, fiber, fuel, medicine, water, and shelter from on-site

How Is It Regenerative?
Reversing our carbon footprints from degenerative to regenerative is key to healing ourselves and our world. Demonstrating a working model of an advanced permaculture homestead is vital in every bioregion and microclimate where people currently live.

Timeline
1-3 years

Benchmarks
1. **Homestead Food Systems Established:** Gardens, Food Forests, and Animal Systems to Match 75-100% or more of Consumption
2. **Homestead Waste Systems Established:** composting systems for all waste, compost toilets, and zero (or near zero) waste systems that either refrain from using wasteful practices or compost or utilize all components onsite
3. **Homestead Rainwater, Graywater, & Blackwater Systems Established** - reed bed and other lagoon graywater and blackwater filtration systems feeding into a food forest and perennial native habitat,
4. **Homestead Power & Fuel Solutions Established**- setup solar panels, wind turbines, water turbines, biogas systems, climate battery (low grade geothermal heating/cooling) and battery systems of all kinds (weights and pumped water can both be used to store potential energy)
5. **Mini-Documentary Created** - Site Tour of the Advanced Permaculture Homestead with a camera and posted on Youtube or other social media publicly available

<u>Sample #3</u>
Mushroom Cultivation & Fungal Compost CSA
Level of Difficulty: LOW-MEDIUM
Level of Benefit: LARGE - MASSIVE

Goal
To start a community supported mushroom cultivation and fungal composting business

How Is It Regenerative?
Mushrooms can be grown on agricultural and other waste substrates to grow edible and medicinal mushrooms and after several flushes create fungal dominant vermicompost that will sequester the carbonaceous exhausted substrate in garden, ranch, orchard, or farm soils.

Timeline
3-6 months

Benchmarks
1. **Regenerative Business Plan & CSA Member Agreement Established**

2. **Mushroom CSA Minimum Membership Achieved**: go to events, restaurants, grocery stores, farm stands, & farmer's markets. Reach out both online and offline. Team up with popular CSA's looking to add variety to their offerings.
3. **Equipment & Supplies Needed Are Acquired**
4. **Business Location Established:** Sign the Lease & Setup Interior & Composting Area
5. **Mushrooms Are Cultivated & Mushroom Products Created & Delivered to CSA Members** & eventually fungal compost

Sample #4
Regenerative Ranching CSA using Leased Land & Mobile Equipment
Level of Difficulty: HARD
Level of Benefit: LARGE

Goals
To regenerate soil, improve the health of the forage plants, and to raise ethical, healthy meat bioregionally using holistic management, portable equipment, and the CSA model.

How Is It Regenerative?
Animals are powerful regenerative forces on the landscape when they are moved regularly and in herd densities similar to how wild herds of herbivores use to roam all grasslands, pursued by large predators. Their grazing, disturbance, density, and movement patterns kept soils, grasslands, herbivores, and carnivores healthy. Using animals, we can sequester carbon, generate ethical, regenerative meat and animal products, and rewild and restore grasslands.

Timeline
3-24 months

Benchmarks
1. **Regenerative Business Plan & CSA Member Agreement Established**
2. **Regenerative Meat CSA Minimum Membership Achieved**: go to events, restaurants, grocery stores, farm stands, & farmer's markets. Reach out both online and offline. Team up with popular CSA's looking to add variety to their offerings.
3. **Equipment & Supplies Needed Are Acquired**
4. **Grazing Land Located & Lease is Signed**: Local processing & CSA Pickup are established.
5. **Herd Is Acquired & Holistic Management Plan & System Established**
6. **Seasonal Delivery of Regenerative Meat to CSA Members is Established**

Sample #5
Regenerative Organic Market Gardening on CSA member land
Level of Difficulty: MEDIUM - HARD
Level of Benefit: LARGE

Goals
To provide regenerative organic fresh produce to local CSA members using CSA member land

How Is This Regenerative?
CSA members can reverse the carbon footprint and degenerative impact of their food consumption by hyper-localizing regenerative no-till or low-till production.

Timeline
2-5 years

Benchmarks
1. **Regenerative Business Plan & CSA Member Agreement Established**
2. **Regenerative Meat CSA Minimum Membership Achieved**: go to events, restaurants, grocery stores, farm stands, your neighbors within walking/riding/driving distance, & farmer's markets. Reach out both online and offline. Team up with popular CSA's looking to add variety to their offerings.
3. **Equipment & Supplies Needed Are Acquired & Local CSA Pickup Location & Schedule Are Established**
4. **Garden Areas & Soils Are Prepared along with Management Plan & Calendar**
5. **The First Season of CSA Crops Are Planted Using Local Regenerative Organic Seeds**
6. **Regular Delivery of Regenerative Produce to CSA Members Is Established**
7. **Organic Certification and then Regenerative Organic Certification Are Established**

<u>Sample #6</u>
North American Western Coastal Restoration, Kelp Farming, & Soil Regeneration CSA Model
Level of Difficulty: HARD - EXTREME
Level of Benefit: MASSIVE - EXTREME

Goals
To restore the coastal regions back to thriving ecosystems, the North American west coast kelp forests, and the soils of farmlands that have eroded into the rivers, streams, oceans, and coastal areas.

How Is This Regenerative?
Kelp can sequester carbon an order of magnitude greater than terrestrial plants – restoring the devastated west coast kelp forests will create a continuous carbon sequestration event as the excess carbon and other nutrients can return to depleted soils instead of causing eutrophication and algae blooms. Shellfish in cages and mussel socks can also be used to filter waters entering coastal areas. Feeding kelp to cattle also dramatically reduces methane amounts released by the cows. The health and food benefits of kelp are also innumerable.

Timeline
2-5 years

Benchmarks
1. **Regenerative Business Plan & CSA Member Agreement Established**
2. **Kelp Farm CSA Minimum Membership Achieved**: go to events, restaurants, grocery stores, farm stands, local health stores, local farmers & ranchers, composting companies, & bioregional health and medicinal product producers. Reach out both online and offline.
3. **Equipment & Supplies Needed Are Acquired & Local CSA Pickup Location & Schedule Are Established**
4. **Kelp and/or Shellfish Farm Site Is Established with Full Permits & Leases**
5. **Kelp and/or Shellfish Farm System Established & First Crop Planted**
6. **First Kelp &/or Shellfish Harvest, Processing, & CSA Delivery**

<u>Sample #7</u>
Community Composting + Urban Food Forest and/or Gardens
Level of Difficulty: MEDIUM - HARD
Level of Benefit: LARGE - MASSIVE

Goals
To channel urban compostable waste streams into a community compost system that builds soils in an urban food forest and/or garden system that feeds local people the most nutritious and fresh food possible.

How Is This Regenerative?
By composting and growing food in what would have been garbage, we are building soil, sequestering carbon, preventing compostable organic matter from entering landfills or being mixed with contaminating waste, providing healthy food for communities, and drawing communities closer together around natural cycles and responsible community practices.

Timeline
1-3 months

Benchmarks
1. **Community Compost Plan, Food Forest/Garden Design & Management Plan, & Compost & U-Pick Food Forest/Garden CSA Member Agreement Established**
2. **Community Compost CSA Minimum Membership Achieved**
3. **Equipment & Supplies Needed Are Acquired & Organic Matter Pickup System & Composting Schedule & Management Team Are Established**
4. **Composting Area & Food Forest/Gardens Are Located, Permitted, & Leased**
5. **Composting Area & Food Forest/Gardens Are Setup & Installed**
6. **First Compost Batch Is Delivered to the Food Forest/Gardens**

Sample #8
Permaculture Orchard - U-Pick CSA
Level of Difficulty: MEDIUM
Level of Benefit: LARGE

Goals
To plant and manage a permaculture orchard designed for CSA members to pick their produce themselves.

How Is This Regenerative?
Permaculture orchards sequester large amounts of carbon, provide habitat, mitigate the climate, create more regular rain, and feed people and all biodiversity in the area the best possible perennial foods.

Timeline
1-5 years

Benchmarks
1. **Land Located & Acquired**
2. **Permaculture Orchard Design & Management Plan Established & CSA Agreement Created**
3. **Minimum # of CSA Members Signed Up**
4. **Equipment, Plants, & Supplies Acquired, Roles Assigned, & Workers Hired (if necessary)**
5. **First Harvest Is Handpicked by CSA Members**

Sample #9
Social Permaculture Project: Create Curriculum, Community Service, Awareness Event, Community Building Event, Church Gardens, School Gardens, & More
Level of Difficulty: LOW - HARD
Level of Benefit: LARGE - MASSIVE

Goals
To share permaculture with the greater world in a way that is empowering, caring for people, and pairing with something they already value and participate in, like school, church, or another community group.

How Is This Regenerative?
By pairing humanitarian service with natural cycles and regenerative skills in a local setting, we can educate local people as we serve them, meeting them where their needs are.

Timeline
1-9 months

Benchmarks
1. Social Permaculture Service, Non-Profit, Community Building, or Business Plan Established
2. Local Community Leader & Organization Relationships Established
3. Site Is Located, Permitted (if applicable), Reserved, & Rented/Leased (if applicable)
4. Site/Event/Service is Promoted by All & Local Community Members Are Reached & Invited
5. Site/Service/Event Setup & Launched

Sample #10
Research/Experiment
For Example: IMOs 1-5 testing in multiple climates on a microscopic level: ID-ing and cataloging before, during, and after fermentation.
Level of Difficulty: MEDIUM - HARD
Level of Benefit: MEDIUM - LARGE

Goals
To further explore or prove an anecdotal or unexplored in-depth achievement using natural cycles and systems to better understand and spread that understanding to others to facilitate greater application and adaptation of methodologies.

How Is This Regenerative?
Testing in multiple climates and contexts doesn't just prove the rule or insights from an anecdotal source, but it creates greater depth of understanding and helps show the parameters and functionality of that method or strategy. This also is likely to reveal weaknesses in universally accepted concepts and ideas, providing critical insights into the complexity that many of these techniques and methods rely upon.

Timeline
1 month–several years

Benchmarks
1. Experiment/Research Plan and Rationale Created
2. Local & Global Regenerative & Scientific Community Leader, Member, & Organization Relationships Established
3. Site Is Located & Leased/Rented/Acquired
4. Site Is Setup for The Experiment/Research
5. 1st Experiments Are Launched
6. Results Are Shared with Peers & The Public

V. The Future

Permaculture is a lens that exponentially adjusts our sight. It opens a window that keeps opening wider, allowing more and more light and hope in. If we want science and all education to be rooted in positivity, regenerative action, inspiration, ethics, and hope, we need to plant our seeds in the soil of permaculture and nourish it with our energy, time, and attention.

We must consciously grow that which does not exist.

We must bring natural beauty and wonder into a world that is rapidly becoming unfamiliar with them. We must help humans reawaken to their human nature. To do that we must break the othering - the idea that we are separate from the world around us, from nature, from animals, from soil, and from each other. We can end this illusion of disconnection by immersing ourselves in the natural world, its systems, its beauty, its richness, its sweetness, and its raw abundance. It is a miracle that we exist - that we ALL exist, from the soil microbe to small child. We live on a very unique and special patch of syntropy - the only one we know of. The Earth and all life stands as a profound and stark reminder of how unique we truly are in the universe - it is up to us to protect, steward, and spread life, no matter the current restrictions, levels of degradation, and costs associated with protecting life: **it is our purpose here on Earth.**

It's time to teach our children the truth of who we are and unlock their full potential. No longer do we need to prepare our children for war, for service in degrading corporations, or for a life of nature deficiency.

The Future is In Your Hands
You Have the Power to Make Real Change
Use this Book & the Related Curriculum to Spread the Regenerative Path
It's All Possible if We Just Invest in the Future.

84

About the Author

Matt Powers is a teacher, author, seed saver, plant breeder, and family guy who teaches people all over the world how to live more regeneratively, so they can have a more abundant and regenerative today and tomorrow through books, online courses, curriculum, and speaking. Over the course of a decade and with a masters degree in Education, Matt went from teaching at a private music school to teaching high school students to teaching high school teachers and administrators to teaching districts and speaking at universities and conferences all over America and online, teaching permaculture and sustainable, regenerative skills and thinking. Matt provides daily inspirational and regenerative content online and is one of the most-followed permaculture teachers online with over 27,000 Twitter followers and tens of thousands of followers in his many Facebook groups and pages ranging in topics from permaculture education to entrepreneurship to gardening to fungi & more.

www.ingramcontent.com/pod-product-compliance
Lightning Source LLC
Chambersburg PA
CBHW051420070526
44584CB00023B/3513